I0484292

United States
Environmental Protection
Agency

Photo courtesy of USDA NRCS.

Protecting Water Resources with
Smart Growth

Acknowledgements

The U.S. Environmental Protection Agency (EPA) extends thanks to the people and organizations who contributed insights and comments on this document as it was being developed: Chester Arnold, University of Connecticut; David Batchelor, Michigan Department of Environmental Quality; Sue Beede, Organization for the Assabet River; Victoria Binetti; Jessica Cogan, Smart Growth Leadership Institute; Thomas Debo, Georgia Institute of Technology; Robert Goo; Joel Haden, Tennessee Valley Authority; Chris Hathaway, Lower Columbia River Estuary Program; John Jacob, Sea Grant Marine Advisory Service, Texas; Jamal Kadri; John Kuriawa, National Oceanic and Atmospheric Administration, Office of Ocean and Coastal Resource Management; Tim Lawrence, Ohio State University Extension; Loretta Lohman, Colorado State University Cooperative Extension; Sylvia Malm; Philip Metzger; Jay Michaels, Minnesota Soil and Erosion Control Agency; Brian Miller, Purdue University, Illinois-Indiana Sea Grant College Program; Scott Millar, Rhode Island Department of Environmental Management; Milt Rhodes, North Carolina Smart Growth Alliance; Evan Richert, Maine State Planning Office; Chris Rowe, Colorado Watershed Network; John Rozum, University of Connecticut; Duane Sands, Iowa Natural Heritage Foundation; Claire Schary; Gene Schiller, Southwest Florida Water Management District; Dave Skellie, Pennsylvania State University Cooperative Extension System; Julie Westerlund, Minnesota Department of Natural Resources; and Don Witherall, Maine Department of Environmental Protection. Additional thanks is extended to EPA staff from the following offices: Office of Water, Office of Ground Water and Drinking Water, EPA Region 1, EPA Region 6, and EPA Region 10. Finally, special thanks are extended to the staff at the Development, Community, and Environment Division and its two interns: Madelyn Carpenter and Marianne Difatta.

ICF Consulting produced parts of this document under EPA contract 68-W-99-054 for the Development, Community, and Environment Division, U.S. EPA Office of Policy, Economics, and Innovation. Eastern Research Group edited and designed the report. EPA staff member Lynn Richards managed this project.

To request additional copies of this report, contact EPA's National Center for Environmental Publications at (513) 891-6561 and ask for publication number EPA 231-R-04-002.

To access this report online, visit <www.epa.gov/smartgrowth> or <www.smartgrowth.org>.

Table of Contents

Photo courtesy of David Cooper.

INTRODUCTION

How are zoning codes and building designs related to standards established by the Clean Water Act? How do transportation choices or the mix of uses in a district affect the quality and quantity of stormwater runoff? How are development patterns associated with protecting the nation's water resources, including lakes, rivers, streams, and aquifers?

As studies have shown, growth and development can have profound effects on our water resources.[1] Storm sewer overflows and polluted runoff from nonpoint sources are a major reason that some water bodies do not meet Clean Water Act (CWA) standards. One factor related to persistent water pollution problems is our development patterns, particularly patterns of highly dispersed development that have been common since the end of World War II. The more woodland, meadowland, and wetland areas disappear under impermeable cover, and the more miles and vehicles we drive and park on impermeable roads and highway surfaces, the more difficult protecting the quality and quantity of our water supplies becomes.

In response to these current trends, local governments are developing smarter approaches to growth. They are looking for, and using, policies and tools that enhance existing neighborhoods, improve schools, protect drinking water, and provide solid housing and transportation choices. These communities are seeking smart growth—a development approach characterized by 10 smart growth principles (see Figure 1). These principles support economic development and jobs; create strong neighborhoods with a range of housing, commercial, and transportation options; and achieve healthy communities and a clean environment.[2]

WANT MORE INFORMATION?

The Smart Growth Network and Smart Growth America have posted information, tools, and resources on all aspects of smart growth on their Web sites: <www.smartgrowth.org> and <www.smartgrowthamerica.org>.

Protecting Water Resources with Smart Growth **is intended for audiences already familiar with smart growth, who now seek specific ideas on how techniques for smarter growth can be used to protect their water resources.** This document is one in a series of publications on smart growth produced by the U.S. Environmental Protection Agency (EPA). Earlier publications, such as EPA's *Our Built and Natural Environments*, or the International City/County Management Association's *Getting to Smart Growth: 100 Policies for Implementation*, *Why Smart Growth*, and *Best Development Practices*, provide basic background on smart growth and a broad range of smart growth techniques.

Smart growth principles provide a foundation—a basic springboard—for the 75 policies described in this report. The majority of these policies (46) are oriented to the watershed, or regional level; the other 29 are targeted at the level of specific development sites.

Section I of this report describes how communities have used smart growth techniques at the regional level to minimize the impacts of new development on their water resources. Communities have been successful by implementing policies to preserve critical regional watershed areas, and strategically directing development to existing communities to minimize runoff from impervious surfaces such as roadways, driveways, and rooftops.

Section II discusses site-level techniques that local governments have used to further mitigate the impacts of development. When used in combination with regional techniques, these site-level techniques can prevent, treat, and store runoff and associated pollutants at the site. Many of these practices incorporate some elements of low-impact development techniques (e.g., rain gardens, biorention areas, and grass swales), although others go further to incorporate smart growth principles such as changing site design practices. Incorporating these techniques can help localities not only to meet their water quality goals, but also to create more interesting and livable communities.

Figure 1: Smart Growth Principles

1. Mix land uses.
2. Take advantage of compact building design.
3. Create a range of housing opportunities and choices.
4. Create walkable neighborhoods.
5. Foster distinctive, attractive communities with a strong sense of place.
6. Preserve open space, farmland, natural beauty, and critical environmental areas.
7. Strengthen and direct development towards existing communities.
8. Provide a variety of transportation choices.
9. Make development decisions predictable, fair, and cost effective.
10. Encourage community and stakeholder collaboration in development decisions.

The examples provided in both sections are drawn from communities across the country. Many policies are supplemented by "practice tips" that illustrate their application or identify additional resources to aid communities with implementation. In addition, several policy descriptions include "issues to consider," which highlight potential complications or other concerns associated with implementing a policy. The experience of local governments has shown that regional and site-specific policies will be most effective when implemented together; addressing the regional or site level alone might not be effective in achieving lasting changes in water quality.

---

[1]*The National Water Quality Inventory: 2000 Report to Congress* identified urban runoff as one of the leading sources of water quality impairment in surface waters. Of the seven pollution source categories listed in the report, "urban runoff/storm sewers" was ranked as the fourth leading source of impairment in rivers, third in lakes, and second in estuaries. See U.S. EPA. *National Water Quality Inventory: 2000 Report to Congress.* <www.epa.gov/305b>. In addition, see Beach, D. 2002. *Coastal Sprawl: The Effects of Urban Design on Aquatic Ecosystems in the United States.* Pew Oceans Commission, Arlington, VA.

[2]U.S. EPA. Development, Community, and Environment Division. April 2001. "What is Smart Growth?" EPA 231-F-01-001A.

Photo courtesy of USDA NRCS.

SECTION I:

Protecting Water Resources at the Regional Level

The amount of land consumed by development has grown dramatically in recent decades, accelerating especially in the 1990s. Between 1954 and 1997, developed land area has almost quadrupled, from 18.6 million acres to about 74 million acres in the contiguous 48 states.[3] In 1997, developed land totaled about seven percent of the nation's nonfederal land area; however, from 1992 to 1997, the national rate of development more than doubled. During this five-year period, more land was developed (nearly 16 million acres) than during 1982 to 1992 (about 13 million acres). The newly developed land came mostly from forestland, pasture and range, and cropland.[4]

The growth of developed areas has led to an increase in impervious surfaces—including rooftops, roads, parking lots, sidewalks, patios, and compacted soil.

Research has shown a strong inverse relationship between impervious cover and water quality. Studies have demonstrated that areas with as little as 10 percent impervious surface[5] within a watershed can impair water resources.[6] Water resources are impacted by activities associated with the construction and use of impervious surfaces. Runoff from the construction of buildings, roads, and sidewalks; emissions generated by travel; and the use of chemicals for landscaping all negatively impact water quality. In addition, byproducts of these activities—such as lawn fertilizers and oil and other waste products from motor vehicles—can combine with runoff and enter stormwater drains, contributing substantially to water pollution. For watersheds, the 10-percent imperviousness threshold can serve as an indicator of the cumulative impacts of these contributing factors.

Although the 10-percent threshold is an indicator of likely impairment at the watershed level, it does not translate well as an indicator at the site level. Some communities have applied the 10-percent threshold figure at the site level, however, with the belief that less imperviousness at the site level will protect water quality. Such applications of the threshold have led some communities to limit population densities to protect water quality. A common approach is the use of zoning to limit housing density to one unit per one, two, or even five acres. This approach attempts to minimize hard surfaces at the site level and therefore preserve absorbent surfaces.

Photo courtesy of USDA NRCS.

This building site in Dallas County, Iowa, represents many low-density development practices.

Low densities at the site level can increase imperviousness at the watershed level, however, leading to worse overall water quality. This effect is due to the fact that the infrastructure and housing footprint requirements for low-density development at the site level can increase the rate at which land within the watershed is developed. As previously undeveloped land is converted to developed uses, pervious open space and naturally absorbent land is converted to roads, houses, shopping malls, businesses, and other uses. The compacted lawns that typically accompany this style of development function much differently than natural green space. In addition, such development also requires greater amounts of transportation-related impervious infrastructure, such as roads, driveways, and parking lots. Finally, if a development is entirely auto-dependent—which is generally the case with low-density development—it can increase vehicle miles traveled and associated air pollution, which also impacts water quality through air-to-water deposition.

On the other hand, smart growth approaches—such as reusing previously developed land; regional clustering; and

WANT MORE INFORMATION?

EPA's Office of Water developed a "Growth and Water Resources" fact sheet that discusses the interaction between development and water quality. It suggests the following guidelines for building communities that protect water resources:

- Establish community goals for water resources in the watershed.
- Direct development where most appropriate for watershed health.
- Minimize adverse impacts of development on watershed health.
- Promote opportunities for restoration.
- Assess and prevent unintended consequences of federal, state, or local decisions affecting watershed health.
- Plan for safe, adequate, and affordable water supplies as an integral part of growth.
- Consider the cumulative impacts of growth management decisions on the watershed.
- Monitor and evaluate the success of initiatives.

More information is available online at: <www.epa.gov/water/yearofcleanwater/docs/growthwater.pdf>.

developing traditional towns, villages, and neighborhood centers—can accommodate the same activity on less land. In turn, this approach reduces overall imperviousness at the watershed level, thus maintaining watershed functions. As stated in EPA's *2003 Draft Report on the Environment*, higher population densities in concentrated areas can reduce water quality impacts from impervious surfaces by accommodating more people and more housing units on less land.[7]

Regional efforts are often needed to effectively coordinate local approaches to development and achieve better watershed-wide results. Regional planning is the process of evaluating potential impacts and formulating approaches for growth in an area that often extends beyond local jurisdictional boundaries. The planning might be carried out by a watershed commission, metropolitan region, county, or other multi-jurisdictional organization. In particular, regional cooperation and planning can be helpful for implementing smart growth approaches such as:

■ Minimizing imperviousness at the watershed level rather than the site level.

■ Identifying and preserving critical ecological areas and contiguous open space areas.

■ Making maximum use of existing infrastructure and previously developed sites.

Successful regional approaches, like those that follow, can reduce overall levels of pollution and still achieve local economic and community goals. The policies featured in the remainder of this section are divided into four issue areas: 1) encouraging development in strategic areas, 2) funding and fee structures, 3) regulatory innovations, and 4) educational efforts.

ENCOURAGING DEVELOPMENT IN STRATEGIC AREAS

Communities should determine areas where they want growth to occur and areas they want to preserve. When such areas are clearly defined, development is encouraged on land with less ecological value, such as previously developed areas (e.g., brownfields, greyfields) and vacant properties. Land with higher ecological value, such as wetlands, marshes, and riparian corridors, is then preserved or otherwise removed from the pool of "developable land."

The policies in this section focus on regional planning practices that can lead to substantial water quality benefits. For the most part, these policies support Smart Growth Principles #6, "Preserve open space, farmland, natural beauty, and critical environmental areas," and #7, "Strengthen and direct development towards existing communities." The policies help communities protect water quality by determining which lands have the highest environmental value, and then establishing provisions to preserve or limit development on those lands. Development is directed to areas identified as most appropriate for growth and where potential runoff impacts can be minimized.

Policy 1. Conduct watershed planning

Watershed planning is a decisionmaking framework that considers water resources and land uses within an entire watershed area (defined by hydrological boundaries) when planning for growth and development. This type of planning allows each watershed to identify specific assets, goals, challenges, and needs that affect the area, yet cross jurisdictional lines. By identifying priority areas for preservation and development at the watershed level, watershed planning helps communities develop policies and incentives to accommodate growth while minimizing impact. Watershed planning requires cooperation from a variety of stakeholders, such as state and local governments, homeowners, environmental organizations, and industry.

Issues to Consider: Managing water resources at the broader watershed scale is likely to require inter-jurisdictional cooperation. Overlapping jurisdictions might require that a new entity be formed to coordinate, manage, and/or enforce the policies generated by the watershed coalition of localities. Such an entity could be invested with advisory authority only, or it might be given authority to enforce watershed-wide policies in member jurisdictions.

Practice Tip: The New Jersey Department of Environmental Protection (NJDEP) recognized that watershed pollutant loads, water withdrawals, and various land uses were creating new management issues that could not be addressed by regulatory programs alone. NJDEP created a watershed management process to address these issues. NJDEP and the New Jersey Water Supply Authority developed a partnership to implement this process in the Raritan River Basin, which provides potable water for nearly 1.2 million people and offers a host of recreational opportunities, habitats for aquatic life, and aesthetic benefits to nearby residents.[8] The goal of this collaborative planning effort was to involve all stakeholders, including farmers, developers, watershed groups, and communities to develop a watershed management plan for the Raritan River Basin. The resulting watershed management plan supports development of target pollutant load reductions, determines where and how development will occur, and identifies specific actions for restoring certain watershed functions.

Policy 2. Develop a regional comprehensive plan

A comprehensive plan (also known as a master plan or general plan) illustrates a community's vision for future growth and development. Most often completed by cities or counties, comprehensive plans project population growth, economic activity, land uses, and other related issues for five-, 10- or 20-year periods into the future. In some cases, states might review the plans to ensure compliance with state guidelines for growth and/or with federal guidelines for certain types of funding.

Comprehensive planning is equally valuable at the regional level, which is typically multi-jurisdictional. Absent a state law to mandate regional comprehensive planning, a region can build a coalition to manage growth by voluntarily establishing a comprehensive planning approach. The state could support the effort by offering incentives to regions to initiate this process. The comprehensive planning process—although sometimes laborious and difficult—can be an effective way for different groups to discuss common regional goals and understand each other's priorities.

Such an effort might be best initiated by focusing on a specific aspect of regional growth, such as an area's natural resources and their relationship to future land use. To achieve water quality goals, comprehensive planning could focus on watersheds and be used to create plans to direct development to encourage sustainability of the region and protection of the region's water resources. Such a method could build on an approach taught by the Nonpoint Education for Municipal Officials (NEMO), which focuses on completing natural resource-based inventories as a type of comprehensive plan. Recognizing that lands have different ecological value, NEMO recommends identifying three categories of land: 1) land that has been developed or otherwise is not in circulation, 2) land that contains critical natural resources that must or should be preserved in perpetuity (e.g., wetlands), and 3) land that is developable (e.g., brownfield, infill, and greyfield sites). This process allows planners to take a larger, regional view of available land and natural resources and combine this knowledge with current development and growth trends. This planning approach, if implemented consistently across the watershed, can produce a realistic, implementable plan to guide development at the regional level.

> **WANT MORE INFORMATION?**
>
> *Natural Resource-Based Planning for Watersheds: A Practical Starter Kit*, a simple booklet that explains NEMO's watershed planning approach, is available online at: <www.nemo.uconn.edu/publications/index.htm>.

NEMO demonstrates its natural resource-based planning for a local watershed group.

Photo courtesy of the NEMO program and the University of Connecticut.

Practice Tip: In August 2000, in Denver, Colorado, five counties and 25 municipalities representing more than 75 percent of the region's population adopted the Mile High Compact, the nation's first voluntary city- and county-led effort to guide growth. The compact is an intergovernmental agreement, through which cities and counties agree to develop comprehensive plans or master plans that support Denver's Metro Vision 2020, the region's long-term plan for growth. Growth consistently ranks in citizen surveys as the region's top concern, and the compact demonstrates that local elected officials are responding to and managing growth. Many mayors, city council members, and county commissioners officially committed themselves and their communities to the process of planning for growth by attending a symbolic signing ceremony to kick off the Mile High Compact. Adopted in 1997, Metro Vision 2020 has six core elements: 1) urban growth areas; 2) a balanced, multi-modal transportation system; 3) preservation of open space; 4) urban centers; 5) free-standing communities; and 6) clean air and water for the region.[9]

Policy 3. Implement watershed-based zoning districts

Local governments are most often the jurisdictions responsible for implementing ordinances or regulations—including zoning—that govern land use. These regulations are sometimes consistent with those of neighboring jurisdictions or with water quality protection principles. Land use planning for water protection is most effective when it covers all land that affects the waterbody in question. Therefore, establishing watershed-based zoning districts can support a comprehensive management approach.

Watershed-based zoning involves defining existing watershed conditions, projecting potential future impervious cover, and redistributing future growth and development through plans and zoning to those areas that would have the least impact on stream or lake water quality. To assist in this effort, zoning districts can be established to set an overall impervious cover threshold or limit for the district. Watershed-based zoning implies that some portions of a watershed will be developed more intensely than others, but the overall goal is to reduce impervious cover. Specifically, a watershed-based zoning approach should include the following steps:

1. Conduct a comprehensive stream inventory.

2. Measure current levels of impervious cover.

3. Verify impervious cover/stream quality relationships.

4. Project future levels of impervious cover.

5. Classify subwatersheds based on stream management "templates" and current impervious cover.

6. Modify master plans/zoning to correspond to subwatershed impervious cover targets and other management strategies identified in subwatershed management templates.

7. Incorporate management priorities from larger watershed management units such as river basins or larger watersheds.

8. Adopt specific watershed protection strategies for each subwatershed.

9. Conduct long-term monitoring to periodically assess watershed status.[10]

Practice Tip: Holliston, Massachusetts, experienced unprecedented growth that began to affect regional water resources and the natural systems that support them. As a result, the Charles River Watershed Association developed an environmental zoning approach with five components:

1. Comprehensive wastewater management planning.

2. Assessment and prioritization of environmental resources and their function, and hydrology.

3. A water budget to meet the town's current and future needs.

4. Stormwater management practices.

5. Land use tools to protect and enhance Holliston's drinking-water resources.

The association is working with the Holliston Bylaw Committee to develop environmental zoning bylaws to protect the town's water resources.[11] Once these bylaws are in place, developers will have increased predictability as to where and how they can develop and what, if any, additional mechanisms need to be implemented to protect the community's water resources.

Brownfield or infill properties, such as this one, are perfect areas to designate as special development districts.

Policy 4. Designate special development districts

Special development districts (also known as special zoning districts) are created to achieve comprehensive planning and urban design objectives within a specified area of a community. The special district allows a community to augment existing zoning regulations (if present) with targeted development criteria to achieve a special, geographically focused goal. The process might also facilitate a more comprehensive approach to zoning in areas where no development regulation formerly existed. Special development zones can offer incentives to encourage development in the targeted area that complies with the district goals. For example, a transit-oriented zoning district might feature compact, mixed-use zoning along key corridors, and offer financial incentives or density bonuses to encourage development that supports a greater use of public transit. Other examples include main street revitalization districts, historic districts, and brownfields targeted for redevelopment.

Special development districts can be used to achieve water goals by encouraging development in targeted areas best able to mitigate potential water quality impacts. For example, to reduce stormwater runoff, a community can use districts to encourage development that incorporates site-level filtration features or to absorb higher-density development that represents a lower per capita imperviousness rate than would be the case on the urban fringe. When special development or zoning districts are successful in absorbing development that would otherwise take place on the urban fringe, the pressure to develop open space is reduced and water quality benefits result. Even those districts that aim primarily to achieve other objectives might yield water quality benefits. For example, transit-oriented districts might reduce vehicle emissions and exhaust deposits that pollute water resources through air deposition.

Issues to Consider: Some local governments use impervious surface zoning districts, which generally set maximum levels for the amount of impervious surface within a zone or, more commonly, on a parcel. For example, no more than 20 percent of a parcel can be covered with impervious surfaces, such as rooftops, driveways, or accessory buildings. Although intended to address overall imperviousness within the watershed, application of maximum levels of imperviousness on a parcel-by-parcel basis through a surface district might not help meet stormwater objectives, and could in fact exacerbate water quality problems, particularly on a watershed scale, by encouraging low-density scattered development. From a smart growth perspective, special districts are best used to achieve water quality improvements by creating incentives for infill or more compact, transit-oriented growth, thereby relieving pressure on open spaces; applying imperviousness limits on a parcel-by-parcel basis can be counterproductive.

Policy 5. Coordinate development and conservation plans

Protecting critical natural resources and planning for future development are often handled as two separate planning processes. For example, a regional environmental authority might be responsible for designating areas for preservation and establishing a plan that reflects those priorities. As a separate effort, a local planning authority might create a plan that describes where and what type of future development will take place. Coordinating these two types of efforts can help protect critical water resources such as wetlands and riparian barriers.

The independence of each process is most evident when planning commissioners face vocal opposition to a proposed development, such as claims that a proposed development will destroy the "last" or the "most productive" wetland in the community. Planning commissioners and their staffs might not have access to the type of ecological information they need to determine the validity of these claims. By ensuring access to information about local plans for growth and regional or state plans for conservation, communities can improve the preservation of sensitive lands and increase predictability within the development process. These two different planning efforts can be shown by overlaying maps to highlight potential conflicts in the two plans and to identify areas of commonality in which local development policies can reinforce regional conservation efforts.

Issues to Consider: Some local governments maintain several development and conservation plans; some could have overlapping geographic areas, but others might not. Identifying these different plans and ensuring that the same areas are analyzed can be a challenge, but well worth doing.

Practice Tip: A handful of communities in Southern California have streamlined the planning process to ensure that local development and regional conservation plans support one another. Orange, Riverside, and San Diego counties integrate special area management, habitat conservation, and local development plans with each other. Overlaying and comparing the plans represented the first step in determining potential areas of agreement and disagreement about where development should be directed, and where land should be preserved. This activity resulted in fruitful negotiations during which developers and landowners set aside areas for development, areas for endangered species habitat, and areas that support critical watershed functions. By better coordinating the two types of plans, urban planners and conservationists are better able to protect water resources and habitat by considering development patterns at the watershed level.[12]

Policy 6. Allow higher densities

A community that allows higher densities can accommodate more housing, business, and commercial uses on a smaller footprint than is possible with lower densities. For example, a community that needs to accommodate 100 houses will disturb 10 acres if the allowed density is 10 units per acre, as compared to 100 acres if the allowed density is one unit per acre. A smaller development footprint means less overall impervious cover and less disturbed land, both of which will better protect regional water resources. In addition, higher densities contribute to more vibrant neighborhoods.

Zoning can be modified to encourage higher densities in existing communities and in greenfield developments. For example, when Montgomery County, Maryland, encouraged higher densities in its greenfield development, the first suburban new urbanist development, Kentlands, was born. More than 2,100 residential units and 2 million square feet of commercial and retail uses were accommodated on 236 acres, resulting in a net density of approximately nine units per acre. This level of density was considerably higher than the surrounding community, which had densities in the range of two to four units per acre. If the same 2,100 residential units were accommodated at two units per acre, the development would have required an additional 814 acres of previously undeveloped land. Preserving large, continuous areas of open space and sensitive ecological areas is critical for maintaining watershed services.

Another way to think about higher densities is to imagine that Manhattan, which accommodates 1.54 million people on 14,720 acres (23 square miles),[13] had been developed not at its current density, but at one or four housing units per acre. At one house per acre, Manhattan would need approximately 1.525 million more acres or an additional 2,283 square miles to accommodate its current population. That is approximately twice the size of Rhode Island. At four houses per acre, Manhattan would need approximately 370,000 more acres or an additional 578 square miles. Higher densities reduce the amount of land developed and, therefore, overall regional impervious cover.

Photo courtesy of U.S. EPA.

King Farm, a development in Montgomery County, Maryland, allowed higher densities in order to preserve open space, such as this riparian buffer.

WANT MORE INFORMATION?

EPA maintains a policy database on innovative zoning ordinances online at: <cfpub.epa.gov/sgpdb/sgdb.cfm>.

Practice Tip: New Jersey's State Plan calls for increasing densities in the state by directing development to existing communities and existing infrastructure. Researchers at Rutgers University analyzed the water quality impacts from current development patterns versus the proposed more compact development pattern. The study found that the proposed development would save 122,000 acres of developable land. This savings translates into significantly less water pollution than current development for all categories of pollutants.[14] The reductions ranged from more than 40 percent for phosphorus and nitrogen to 10 percent for lead. Moreover, the proposed development would reduce runoff by 30 percent.[15] These conclusions supported findings from a similar statewide study, completed in 1992, that concluded that compact development would result in 30 percent less runoff and 40 percent less water pollution than a sprawl scenario would.[16]

Policy 7. Use density averaging

Calculating density requires a community to consider the appropriate level of development for a particular area, given the character, neighborhood context, amenities, and anticipated use of the area. In previous efforts to limit impervious cover, some communities have lowered their desired density, thereby dispersing the same amount of development across a wider geographic area. As a better alternative, jurisdictions can use "density averaging" when setting acceptable limits of development, thereby targeting growth to some areas and away from other areas.

Density averaging aids in the preservation of critical ecological areas by helping to direct growth pressures elsewhere. It can provide an option for communities wanting to increase densities in central areas, for example, while limiting growth in more outlying areas. In fact, this practice acts as an informal trading system within a watershed, redirecting growth to areas that can best absorb it and away from areas where it poses the greatest threat. Density averaging allows for the most efficient use of space within areas that might be the most expensive as well.

Density averaging calculates the number of units that could be constructed on a parcel based on existing zoning policies and transfers all or some portion of those units to a separate, non-contiguous parcel.

WANT MORE INFORMATION?

North Carolina encourages density averaging through its watershed management efforts. For more information and a copy of the guidelines, contact the Water Quality Committee of the Environmental Management Commission at (919) 773-5083, ext. 566.

Policy 8. Preserve open space, including critical environmental areas

WANT MORE INFORMATION?

The Center for Watershed Protection offers additional water quality information on critical environmental areas at: <www.cwp.org/aquatic_buffers.htm> or <www.stormwatercenter.net>.

Planning for growth requires that land be identified to accommodate residential, commercial, and industrial needs. In addition, regional planning efforts must consider community needs for open space that provide recreational, aesthetic, and natural functions. An open landscape helps preserve the geographical distinction of an area, thereby fostering a strong sense of place. Indeed, preserving open space is considered so central to successful communities pursuing smart growth strategies that it is listed as the sixth Smart Growth Principle, "Preserve open space, farmland, natural beauty, and critical environmental areas."

Preserving open space is critical to maintaining water quality at the regional level. Large, continuous areas of open space reduce and slow runoff, absorb sediments, serve as flood control, and help maintain aquatic communities. In most regions, open space comprises significant portions of a watershed, filtering out trash, debris, and chemical pollutants before they enter a community's water system. Open space provides a number of other benefits, including habitat for plants and animals, recreational opportunities, forest and ranch land, places of natural beauty, and important community space.

In addition, preserving land that serves strategic ecological functions (e.g., wetlands, buffer zones, riparian corridors, floodplains) is critical for regional water quality. For example, buffer strips decrease the amount of pollution entering the water system. Tree and shrub roots hold riverbanks in place, preventing erosion and resulting sedimentation and turbidity. River and lakeside grasses slow the flow of runoff, giving the sediment time to settle and water time to percolate, filter through the soil, and recharge underlying groundwater. Wooded buffers offer the greatest protection; for example, according to one study, when soil conditions are ideal, a 20- to 30-foot-wide strip of woodland can remove 90 percent of nitrates.[17] By slowing and holding water, wetlands and buffer zones increase groundwater recharge, directly reducing the potential for flooding.

Preserving and maintaining riparian buffer areas are critical for ensuring water quality.

Photo courtesy of USDA NRCS.

Communities are developing open space conservation programs that target the most critical areas for preservation, working with public or nonprofit organizations to acquire lands outright, purchase them, or arrange for conservation easements, which restrict future development. Conservation easements, for example, provide a more economical means to preserve open space than an outright purchase. A conservation easement is a legally binding agreement that limits or prohibits certain uses from occurring on a property that would interfere with its conservation. Although they restrict development, such voluntary easements often allow for land uses—such as limited forest harvesting, limited residential development, or agriculture—that yield financial returns to the property owner and are consistent with the long-term health of the watershed.

Issues to Consider: Acquisition of land or permanent limits on development can be costly, particularly if the targeted open space is in an area under growing development pressure. Privately owned farmland, forests, and other "green space" often represent the lifetime savings of family farmers. As such, any successful attempt to acquire the lands for future public benefit will require an expenditure that closely matches its market value for development. Further, although the community benefits from such open space are numerous, they tend not to be widely understood by the general public, so public outreach and education might be necessary.

Land preservation efforts must be conducted in a comprehensive and consistent manner to ensure that the most critical environmental areas are preserved in their entirety and connected to other areas through greenways or riparian corridors, as appropriate. Without taking a comprehensive approach, land preservation can occur in a scattered manner, effectively eliminating or significantly reducing natural ecological functions. Finally, efforts to preserve land in some areas must also correspond with plans to accommodate development in other areas, ensuring that overall growth is not restricted, but redirected.

Practice Tip: In the face of declining water quality, Hillsborough County, Florida,[18] decided to take a proactive approach to managing development. With the help of Duany Plater-Zyberk & Company, Hillsborough town planners mapped out areas that were currently developed by "blacking out" those areas on a county map; they likewise marked currently preserved or protected areas. County planners, local officials, and citizens then discussed the use of the remaining areas through a series of public meetings and visioning sessions. The process resulted in the identification of additional areas to preserve because of their environmental value, such as riparian buffers and wetlands, or because of their social or recreational value. As a result of this collaborative process, Hillsborough dramatically increased the amount of its open space, thereby better protecting its water resources, and increased predictability for developers, who now have a much better sense of which lands are ripe for future development.

WANT MORE INFORMATION?

The Trust for Public Land developed *Greenprints for Growth*, a step-by-step guide for identifying, purchasing, and managing community open space. It is available through its Web site at: <www.tpl.org>.

Policy 9. Direct development through transferable development rights

Existing zoning, in most cases, prescribes the type and quantity of use that is allowable on a given piece of land. There are few exceptions, such as variances and spot zoning changes, that are likely to result in a different use than that predetermined in the zoning code. A transfer of development rights (TDR) program offers property owners more flexibility in how they use their land and provides communities with a means to redirect growth away from areas most likely to impact a region's water quality.

A TDR program allows landholders in sensitive areas to transfer their development rights to other, more appropriate locations, such as less sensitive areas, or areas where infrastructure already exists. TDR ordinances establish a sending (or preservation) area and a receiving (high-density growth) area. Landowners in the sending area receive credits equivalent to their development rights under current zoning guidelines. They can then sell these credits in exchange for not developing their land (administered through deed restrictions on the sending area parcels) or developing it at a far lower density (administered through zoning restrictions). Real estate developers can purchase these development-right credits and use them to increase existing or planned densities on parcels in receiving areas. By providing an economic incentive for preserving undeveloped land, TDRs allow a community to preserve important open space resources while permitting owners of property in targeted areas to recoup the value of the property's development potential.

Issues to Consider: Some states do not have legislation in place to support such transfers. If the statutory authority does not exist, the aid of state legislators will be required to create an appropriate legislative environment to support the development of local TDR programs.

Practice Tip: In 1980, Montgomery County, Maryland, downzoned agricultural land from a maximum density of one house per five acres to one house per 25 acres. The county also designated this land as a Rural Density Transfer Zone (the TDR sending area), allowing landowners to sell one development right per five acres. The county established an initial receiving area, which could accommodate up to 3,000 development rights. Each development-right purchase entitled receiving area landowners to build one more housing unit per acre than otherwise would have been allowed. By the end of the 1997 fiscal year, the program had accommodated the same amount of overall units, but protected 39,180 acres that would have otherwise been developed. By transferring rights to develop, Montgomery County directed development to previously developed, more appropriate areas, and protected areas that could be more sensitive to development or likely to impact water quality.[19]

Policy 10. Coordinate development planning with sewer and water authorities

Often plans for water and sewer service expansion are more heavily influenced by utilities' projections for future demand than by a community's growth priorities. Once water and sewer expansions are approved and constructed, development frequently follows, whether or not it supports other community goals for targeted and directed development.

Sewer and water authorities can play a major role in directing a region's growth by determining when and where new infrastructure investment will occur. Well-drafted facility planning areas can direct growth by providing sewer service in areas least likely to impact water resources. Decisions on how and where to provide sewer service, as described in a facility planning area, affect not only the quality of wastewater treatment available to residents but also where open land can be developed. Planning/infrastructure coordination is easier if extensions of existing facility planning areas require the approval of the regional or state environmental agency or planning agency.[20] In this way, facility planning areas can be a strong tool to determine how and where a community will grow.

For example, the state of Wisconsin uses planned sewer service areas as a tool to integrate wastewater infrastructure and local planning efforts. As a rule, Wisconsin automatically excludes environmentally sensitive areas such as wetlands, steep slopes, and floodplains from consideration for current or future service extensions. The development of these areas must correspond with the goals of the local comprehensive plan, and not depart from any other ordinances directing growth and resource protection. The state estimates that these efforts to protect natural areas and incorporate land use planning can prevent the loss of millions of dollars due to the destruction of habitats, impairment of water quality, and cleanup associated with failing wastewater treatment methods.[21]

Issues to Consider: Critics of planned sewer areas argue that by directing growth towards designated communities and regions, sprawl and degraded water quality can result. Others dispute the potential role of the state or regional agency in considering local plans for growth. In addition, facility planning areas can cause neighboring municipalities to argue about the placement of sewer service in an effort to attract growth to their own jurisdictions and boost property taxes and other revenues. These conflicts must be addressed and resolved to achieve the maximum beneficial results planned sewer service can provide.

Practice Tip: In Ohio, the city of Columbus' Division of Sewerage and Drainage plays a critical role in shaping the growth of the region. The division has developed a facilities plan update that calls for centralized wastewater treatment services to be provided within the facility planning area boundary. The city will not extend its sewerage services beyond this boundary—strongly encouraging development within the boundary. In addition, recognizing the role that sewer infrastructure plays in regional growth patterns, the facilities plan articulates the following goals[22]:

- Protect critical water resources, especially in the Darby Watershed
- Maximize existing infrastructure investments
- Incorporate watershed planning
- Mitigate stormwater impacts from urban development
- Curb urban sprawl

Policy 11. Limit development on land near public wells

Traditional zoning practices often do not take into account the location of drinking water sources, and as a result might permit growth near public wells. This practice can impact the quality and supply of drinking water sources. Fertilizers, for example, when used on agricultural lands or sites with extensive landscaping (e.g., golf courses) can mix with runoff water and contaminate groundwater sources. Most zoning practices focus on the designated use of a zoned area and do not consider the location of drinking water sources or the impacts development can have on these sources.

Some municipalities have chosen to restrict or prohibit development near drinking water sources using approaches such as zoning or ordinances. Others require the use of best management practices to limit water quality impacts. Limiting development near public wells helps direct development to existing communities, including infill and brownfield sites.

Issues to Consider: Limiting development near drinking water sources can be controversial and require collaboration with potential developers and other stakeholders. Water quality ordinances can help provide flexibility for developers willing to take adequate measures to protect water resources. Unfortunately, limiting development near wells will not completely prevent contamination of groundwater. Contaminants can enter groundwater at areas distant from the wells, particularly in recharge areas, and travel with groundwater flow.

Policy 12. Consider the cumulative and secondary impacts of development in the floodplain

Most state and local governments require only existing development to be included on floodplain maps; however, these maps should also include future development and infrastructure in and out of the floodplain to ensure that floodplains continue to serve their natural functions.

The Federal Emergency Management Agency (FEMA) requires local governments to delineate floodplains. In most cases, designated floodplains are subject to local or federal development restrictions, which can range from requiring flood insurance to incorporating flood mitigation measures. Although local governments must examine current and future development in the floodplain, they do not always consider secondary impacts from that development. For example, local governments might not evaluate future residential development stemming from a new road, but the cumulative impacts of these secondary impacts can be significant, such as increased runoff and peak flow rates from the increase in impervious cover—both of which can expand the floodplain. In Charlotte, North Carolina, for example, impervious surfaces such as parking lots and roads have made it more difficult for water to be absorbed into the ground and, in turn, have expanded the 100-year flood area.

Not considering the cumulative and secondary impacts of development can have disastrous consequences.

To better protect regional water resources, the cumulative and secondary impacts of development in the floodplain should be considered before development occurs. By better representing where and how future development will occur, and incorporating these findings into flood zone maps, communities can ensure that growth is directed away from environmentally sensitive areas where the floodplain areas could be impacted by development, thereby protecting water resources. Communities can further prevent development in flood-prone areas by directing growth to less hazard-prone, more highly developed areas.

Practice Tip: To improve its ability to identify flood-prone areas, and avoid a repeat of the devastating effects of Hurricane Floyd, the state of North Carolina revamped its process of developing floodplain maps by expanding the areas to be included. Charlotte became the first community in the country to include future development on its floodplain maps. The new maps are incorporated into local decisions about where to allow construction, and are used to enforce more stringent regulations for growth in and out of the floodplains. Furthermore, new construction in a regulated floodplain requires a special permit. Charlotte estimates that the new maps will keep more than 1,500 new structures out of the floodplain during the next 30 years, saving Charlotte citizens more than $330 million in possible losses.[23]

Policy 13. Update combined sewer and sanitary sewer systems in downtown areas

Outdated water and wastewater systems can limit development or redevelopment in some areas of the United States. To encourage development in these areas, municipalities and states are advised to upgrade and expand the sewer and water infrastructure in existing communities. For those areas where systems are at or near capacity, but where the municipality still wants to direct development to them for planning reasons, a matching funds program could be made available to developers to mitigate the high costs of sewage repair and replacement.

Expenditures to correct overflow problems and address other lagging maintenance and repair issues could be targeted to redevelopment areas to help revive urban economic vitality, especially in cities that are restoring waterfronts as part of downtown revitalization efforts. Public expenditures on infrastructure, such as streets, highways, water and sewer systems, lighting, and schools and other civic buildings, constitute a significant share of public expenditures each year. Whether they intend to or not, local and state governments are essentially defining locational priorities for new development when they allow infrastructure in existing neighborhoods to decay while investing in new infrastructure in edge communities. By not addressing problems with the older infrastructure, the local government creates a larger fiscal problem each year that the maintenance issues are not evaluated.

Issues to Consider: System retrofits can be costly and can result in increased rates. Higher rates can deter businesses that otherwise would have developed in the city and lead them to relocate in areas where rates are lower.

Practice Tip: In Richmond, Virginia, combined sewer overflows (CSOs) were creating an unsightly and smelly environment that was inhibiting the redevelopment and orientation of tourism surrounding the James River. The city decided to address the CSOs as an aesthetic and environmental problem affecting the city's waterfront. To help solve the city's CSO problem, the Virginia Department of Public Utilities embarked on a $117 million CSO control program. The city identified overflow discharge points in recreational and other public areas and redirected flow to a retention basin. This program also corresponded with the restoration of the historic canals and revitalization of the downtown riverfront.[24] Now the city can promote a more visible riverfront as a civic amenity.

Policy 14. Develop infill sites

Numerous sites in cities across the United States remain underutilized or vacant; in some communities, the number of such properties is growing. For example, in Philadelphia, Pennsylvania, an average of 1,348 properties were abandoned each year from 1984 to 2000.[25] Abandoned properties decrease the value of surrounding properties, pose fire hazards, and attract crime. Therefore, the redevelopment of these properties not only helps revitalize existing communities, but also serves to reduce development pressure on land critical to maintaining water quality.

Infill development means reusing underutilized or vacant land located in an existing neighborhood. Infill development promotes water quality by accommodating growth on sites that could already be impervious, thus eliminating the need for any new impervious cover and the need to disturb new land during construction. Developing infill sites can reduce pressure for development on open land providing critical water functions (such as infiltration or source water supply) on the urban fringe. When they are redeveloped at higher densities, infill sites also provide local governments an opportunity to ensure that more people are located in areas with existing infrastructure, housing choices, and transportation choices.

Infill development might also represent an underutilized resource for communities that otherwise feel that new growth and development can be accommodated only on undeveloped land at the urban fringe. A recent analysis completed by King County, Washington, for example, demonstrated that vacant property eligible for redevelopment in the county's growth areas could accommodate 263,000 new housing units— enough for 500,000 people.[26] Redeveloping these assets represents a significant opportunity for new growth without degrading water quality. Additional onsite landscaping methods or development techniques that mimic the predevelopment site hydrology can further promote water quality benefits. Communities can encourage infill development through funding incentives or flexible regulations and zoning.

Practice Tip: Clark County, Washington, adopted an ordinance in 2002 that encourages infill development and recognizes the stormwater benefits associated with it. The ordinance applies to selected districts as well as lots less than two acres in size that adjoin existing development and are served by existing infrastructure. Two types of infill development are allowed: 1) detached single-family housing with lot sizes smaller than under regular zoning, and 2) attached and detached single-family housing, duplexes, and multi-family housing. Lot coverage can be up to 60 percent, or 70 percent with a variance. Developers might also receive density bonuses, plus infill projects are exempt from stormwater regulations if they create less than 5,000 square feet of new impervious surface.[27]

WANT MORE INFORMATION?

Smart Growth America, the International City/County Management Association, the National Trust for Historic Preservation, and the Local Initiatives Support Corporation recently launched the National Vacant Properties Campaign. Details are available at: <www. vacantproperties. org>.

Policy 15. Redevelop brownfields

Brownfields are abandoned, idled, or underused industrial and commercial facilities where expansion or redevelopment is complicated by real or perceived environmental contamination. Although brownfields can serve as valuable infill development opportunities, many communities require additional mitigation measures, such as best management practices or upgraded sewer pipes, to redevelop an urban brownfield site. Unfortunately, these additional measures can serve as a deterrent to redevelopment.

When brownfield sites are reused, not only is their former environmental threat removed (e.g., leaking oil tanks from previous industrial uses), but their redevelopment can yield other environmental benefits. By absorbing development that would otherwise be directed to greenfield sites on the urban fringe, brownfield redevelopment helps preserve open space located elsewhere in the region. A recent George Washington University study found that for every acre of brownfield that is redeveloped, more than four acres of open space are preserved.[28] In addition, the redevelopment of brownfield sites can also be used to treat, store, and manage stormwater runoff. For example, the cleanup of a brownfield site might not be sufficient for residential development, but it could be clean enough for stormwater runoff mitigation measures, such as creating rain gardens or large grass swales. Using a brownfield site in this manner can also provide habitat opportunities for birds and other species.

In many cases, redevelopment activities can utilize the existing infrastructure that surrounds the site. Residential and commercial development that would have otherwise required new impervious roads and parking lots to be constructed to service the new site can instead use existing resources, thus reducing the overall level of a community's imperviousness.

Brownfield redevelopment can be encouraged through tax or other financial incentives, or regulatory incentives. For example, by recognizing the water quality benefits that brownfield redevelopment provides, communities could apply less stringent runoff standards that would reduce the required level of stormwater runoff mitigation measures, making the project less costly for the developer. This policy does not suggest any loss in water quality for the community, but instead recognizes the importance of development location for water benefits. Several communities have already adopted or are considering this approach. Fairfax County, Virginia, is one community that has reduced the runoff requirements for redevelopment of existing properties.[29] Oshkosh, Wisconsin, is considering such an approach after observing the high cost of compliance with newer, more stringent stormwater regulations for properties in redevelopment areas. Jackson Kinney, Oshkosh's director of community

WANT MORE INFORMATION?

EPA information on how to remediate, market, and develop brownfields is available at: <www.epa.gov/swerosps/bf/index.html>.

New apartments and an urban park replace an industrial brownfield in the Pearl District, Portland, Oregon.

Photo courtesy of U.S. EPA.

development, noted, "In a redevelopment site [in contrast to a greenfield development] you're really not changing the stormwater drainage dynamics from what previously existed."[30]

Practice Tip: Atlantic Station, a redevelopment project on a former industrial site in Atlanta, Georgia, illustrates the water quality benefits achieved by redeveloping a brownfield site. The centrally located, mixed-use site design accommodates approximately 3,100 residential units and 7 million square feet of retail, office, and hotel space on 138 acres. If these same housing units and commercial space were constructed elsewhere in the region at densities typical for the region, the same development would require almost 1,200 acres. By using less land for development, less soil is disturbed during construction, decreasing soil erosion. In addition, modeling results suggest that the compact nature of the Atlantic Station site generates approximately five times less runoff, four times fewer total suspended solids, six times less total nitrogen, and 16 times less total phosphorus than the low-density alternative site designs.[31]

Policy 16. Redevelop greyfields

Greyfield sites are abandoned, obsolete, or underutilized properties, such as regional shopping malls and strip retail developments. Although typically not viewed by communities as potential sites for residential land uses, these properties often have significant redevelopment potential because of their large size, existing infrastructure, and established community presence.

Like other infill development, greyfield redevelopment can absorb growth that might otherwise convert green space on the urban fringe. Redeveloping greyfield properties provides a range of economic and social benefits, including the opportunity to bring new life to blighted commercial spaces, locate new services and amenities in close proximity to existing transit networks, and maximize a community's existing investments in water, sewer, and road infrastructure.

Communities can reap many benefits by converting these large, vacant, or underused shopping areas into new mixed-use neighborhoods. Also, by incorporating smart growth features—such as compact development, the provision of open space, and reduced parking requirements—greyfield redevelopment can yield significant environmental benefits. For instance, redevelopment can actually reduce a site's impervious

Market Common, a development in Arlington, Virginia, that includes stores, apartments, townhomes, single-family houses, parking garages, and a one-acre public park, was built on a former Sears store and parking lot.

Photo courtesy of U.S. EPA.

cover by converting parking areas to pocket parks or buffer zones. By allowing for some natural stormwater infiltration, the site's net stormwater runoff is decreased. Finally, as with brownfield redevelopment, greyfield redevelopment reduces development pressures at the urban edge. Many of the same techniques used to encourage brownfield redevelopment can be applied to greyfield redevelopment as well.

Practice Tip: Redeveloped on the site of a former shopping mall, Mizner Park in Boca Raton, Florida, is an example of a greyfield reborn as a mixed-use development. Redesigned from its original pattern of a large retail structure surrounded by surface parking lots, the 29-acre site now includes 272 apartments and townhouses, 103,000 square feet of office space, and 156,000 square feet of retail space. Most parking is accommodated in four multistory parking garages. Designed as a village within a city, the project has a density five times higher than the rest of the city and a mix of large and small retailers, restaurants, and entertainment venues.[32] More significantly, the final buildout of Mizner Park decreases overall impervious surface by 15 percent compared to the former shopping mall, through the addition of a central park plaza, flower and tree planters, and a large public amphitheater.

Policy 17. Maximize transportation choices

The range and quality of transportation choices available to people not only have a direct impact on where and what type of development is likely to occur, they also exert both an indirect and direct effect on water quality. Where motor vehicle travel is the only practical form of transportation, little incentive exists to depart from the conventional, low-density development designed to accommodate vehicular traffic.

Furthermore, air emissions from vehicles, through air-to-water deposition of pollutants, are a major contributor to poor water quality and can undermine other efforts to improve the quality of a region's water. For example, in the metropolitan Washington, D.C. region, mobile sources (e.g., cars, buses, trucks) are a primary cause of harmful ground-level ozone.[33] The resulting smog not only affects air quality, but it also compromises water quality as pollutants end up in water bodies, through deposition or stormwater runoff. For example, EPA estimates that 35 percent of the nitrogen entering the Chesapeake Bay is from mobile sources.[34] Increasing the viability of alternative transportation can decrease air deposition of pollutants into water resources.

WANT MORE INFORMATION?

In 2002, the Congress for the New Urbanism published *Turning Greyfields into Goldfields: Dead Malls Become Living Neighborhoods*, available for purchase online at: <store.yahoo.com/cnuinfo/greyingoldea.html>.

Well designed sidewalks and roads provide opportunities for walking, biking, driving, and transit.

Photo courtesy of U.S. EPA.

Communities can employ a range of strategies to maximize transportation choices. For example, they can use zoning and tax incentives to create more walkable communities, characterized by mixed land uses, compact building, and inviting pedestrian corridors. Local governments can zone for a mix of uses to develop stores, schools, and restaurants within walking distance of each other. Or they could provide tax incentives to encourage residential development in downtown areas that are dominated by offices.

By providing people choices on how to get to the places they want to go, such as fast, reliable buses and trains; bike paths; or walking routes, air emissions from mobile sources can be reduced. By providing amenities such as bus shelters and bike racks, governments can increase the likelihood that the public will use these alternative transportation methods. Allowing individuals to substitute walking, bicycling, or other modes of transportation for trips that once required a car can reduce congestion and traffic and improve water and air quality.

FUNDING AND FEE STRUCTURES

Monetary incentives and disincentives are powerful tools for influencing, directing, or altering growth patterns to minimize their water quality impact. Fees can be structured to encourage desired outcomes, such as better stormwater control. Fees can be calculated to reflect the true cost of water degradation resulting from development. Plus, fees and service charges are among the most direct means available to communities to demonstrate development and environmental priorities.

The appropriate and even-handed use of fees can augment public loans and other funding resources and provide a needed source of capital for communities to invest in upgrades, expansions, and other enhancements to their water infrastructure systems. Low-interest loans, grants, and other resources are available through federal and state governments, as well as some private and nonprofit sector partners, to help communities improve their water systems through smart growth approaches. This subsection provides examples of funding sources that can be used to improve water quality through smart growth.

WANT MORE INFORMATION?

EPA developed the *Smart Growth Funding Resource Guide*, a list of funding resources for local and state governments, communities, and non-governmental organizations that are addressing the varied aspects of smart growth. It is available at: <www.epa.gov/smartgrowth>.

Policy 18. Create a stormwater utility

Fees to address stormwater issues are generally raised by a local utility through permit fees, water and sewer fees, and any fines levied against a permit violator. These local utilities generally use the funds raised within the locality to address problems or issues in that same area.

WANT MORE INFORMATION?

The Center for Urban Policy and the Environment at Indiana University-Purdue University Indianapolis has extensive information on how to create and manage stormwater utilities. Details are available at: <stormwaterfinance.urbancenter.iupui.edu>.

A stormwater utility, however, is a special district created to generate a stable funding source for stormwater management across a specific region. Funds are generated through user fees and generally used for system upgrades or other stormwater runoff mitigation efforts. The stormwater utility approach provides revenue and a flexible means of implementation applicable under many different state laws and across environmentally diverse areas. Stormwater utilities can also motivate partnerships and support a more regional approach to at least one aspect of water resource management.

Various methods are currently used by stormwater utilities to determine user fees. Many base their fees at least in part on the percentage of impervious cover of developed land, sometimes at the parcel level. Many use the parcel-level calculation only for commercial properties, however, and simply charge a flat rate for residential properties. Some municipalities employ more sophisticated residential user fee calculations that also consider fees for nearby public roads.

As of late 2000, more than 400 municipalities nationwide had created some form of stormwater management utility.[35] Although this approach is still being shaped and gaining momentum, it offers a way to create incentives for smart growth developments, especially as user fee calculation increases in sophistication. For example, waivers or fee reductions could be given for compact construction and high-density development.

Issues to Consider: Calculating utility fees can be challenging and contentious. Many cities have determined that the most equitable approach to calculating utility fees is based on the amount of impervious area on each property. Other factors, such as property size, can also be considered in determining fees. Other communities have determined that a more convenient means to assess fees is to charge a flat rate. Although charging a flat rate could be more cost-effective in the shortrun for residential properties, doing so fails to reflect the benefits of compact, mixed-use development, and thus encourages dispersed, detached development. A compounding factor is that the user fee amount is usually quite small (e.g., $2.95 per house); therefore, it is unlikely to drive alternative site design choices, either by homebuyers or developers. Stormwater utilities typically require enabling legislation at the state level (if statutory authority does not already exist) to be established, as they levy taxes to finance operations and capital improvements.

Policy 19. Use wastewater fees to fund watershed-level planning

Wastewater fees are typically used for wastewater treatment, capacity upgrades, and ongoing operation and maintenance costs. Some communities might also want to consider using some portion of wastewater fees to fund regional watershed-level planning, particularly if lack of funding is an obstacle to watershed-level planning.

Most municipalities have only enough resources to address planning issues within their own jurisdictions. Using a portion of the wastewater fees to support watershed-wide planning will not only support cross-jurisdictional planning, but might also help create a coalition of interested parties. Fees collected through wastewater assessments can be used to fund partnerships or authorities involved in watershed planning. In particular, resources can be used to support pilot projects, technological innovations, infrastructure improvements, or the planning for development in and around a watershed.

Practice Tip: The Cherry Creek Basin Water Quality Authority, a regional water authority created by Colorado's legislature in 1985, operates under state law to undertake various water quality and capital projects and assess fees for the Cherry Creek basin. The Authority is funded through a portion of local wastewater treatment fees (approximately $1.5 million per year) assessed and collected by the authority.[36]

Policy 20. Vary sewer hookup fees for existing and suburban fringe locations

In most communities, sewer hookup fees are calculated and assessed by localities without regard to location, so the same fee applies in suburban as well as central city locations. A more strategic approach is to vary hookup fees by site location, reflecting the distance-dependent costs associated with sewer service and encouraging development in central locations.

Developments like Metro Square in Sacramento, California, are eligible for reduced sewer hookup fees because of their high density and central location.

Photo courtesy of Local Government Commission.

Many municipalities assess the cost of sewer hookup fees on an average-cost basis, which fails to reflect the true cost of system expansion and can serve to support dispersed, low-density development. Conventional approaches to hookup fee assessments treat all new developments equally, regardless of location, compactness, or dispersion. To further direct development and encourage infill, municipalities should consider assessing variable rates for sewer hookup based on location, charging lower hookup fees where growth is to be encouraged, or incorporating design elements that improve water quality impacts in new projects.

Practice Tip: In Sacramento, California, regional sewer officials recently approved plans to dramatically reduce sewer hookup fees in existing neighborhoods and raise fees on the urban fringe. The change is part of a series of planned rate hikes needed to finance a $1.3 billion network of large new pipelines necessitated by rapid suburban growth. It is the first time in the Sacramento Regional County Sanitation District's 25-year history that different rates will be charged based on location. Previously, the district operated on the principle that everyone would pay the same amount to hook up to the sewer system, regardless of location. Under the new plan, the connection fee for a house in a new neighborhood is $5,255; on the other hand, the fee for a new house in an existing urban area is $2,314. Commercial fees are handled the same way. The plan received endorsements from a wide array of supporters, ranging from the Sierra Club to the Sacramento County Taxpayers League.[37]

Policy 21. Direct infrastructure spending to designated growth areas

State and local governments often use infrastructure funding in accordance with multi-year capital investment plans that determine priority areas for growth and construction, among other needs. State and local governments can direct infrastructure spending to designated growth areas in existing communities as one way of encouraging development activity in areas where private and public investments have already occurred.

Across the country, water and sewer infrastructure is aging, and municipalities are faced with choices on where—and according to what priorities—to invest in their water and sewer infrastructure. Their allocations could be based on projected tax revenues from new development supported by the current infrastructure or on infrastructure in greatest need of repair. Strategically targeting infrastructure resources to direct development to designated growth areas in existing communities is another approach for prioritization. For example, the state of Maryland created its "priority

WANT MORE INFORMATION?

The state of Maryland provides information on its priority funding area program, including models and guidelines, online at: <www.mdp.state.md.us/smartgrowth/pdf/PFA.PDF>.

funding areas" effort in 1997. Since then, the state has provided infrastructure funds for roads, sewer, water, and schools only in those communities targeted for new development based on their existing resources, such as transit facilities, infrastructure, or infill opportunities. Any development that occurs outside the priority funding area does not receive state financial support.

Other states have prioritized the use of infrastructure funds for repair and maintenance before funding new construction—another way to direct infrastructure funds to existing areas designated for future growth. For example, in 2002, New Jersey announced its Smart Growth Infrastructure Tax Credit program, directing limited state resources to support areas with existing infrastructure. This $10 million program will provide tax incentives to encourage builders and developers to invest in neighborhoods that have existing or already planned infrastructure. Administered by the New Jersey Housing and Mortgage Finance Agency in consultation with the State Planning Commission, the program offers tax credits to eligible residential, commercial, and mixed-use retail projects. Such projects are developments located in municipal (urban) aid areas, municipalities with designated centers, or municipalities with plans endorsed by the State Planning Commission.[38]

Practice Tip: Vermont's Agency of Natural Resources encourages communities to direct growth to downtown and other planned growth centers, while managing growth in the surrounding countryside. The agency gives priority to ensure that older, failing wastewater treatment facilities receive needed improvements, rather than directing resources to newer plants that would support development on the urban fringe. The agency is currently revising its rules to implement this "fix-it-first" approach to help communities consider the relationship between infrastructure planning and land use planning during the earliest project planning stages, thereby avoiding permitting conflicts.[39]

Policy 22. Differentiate development fees based on location of the development

Studies have shown that infrastructure costs increase when development takes place beyond the local service area.[40] The higher costs incurred are due to the necessity of providing longer trunk lines and connecting roads for more distant and dispersed development. These costs tend to increase based on the distance from the urban core and from other housing units. Although they have generally not done so in the past, local governments have the option of charging fringe-area developers the full costs for providing infrastructure.

Some localities assess developers only partial fees for infrastructure costs for services such as water, sewer, roads, and schools.[41] They do this in order to attract development. This is a costly practice, however, because new residential development costs municipalities more than the revenue it generates. The negative impact on local government budgets is often not readily apparent because of the timing of evaluating the actual costs and revenues. Early on, during construction, building activity provides attractive tax revenues to the local government. At the same time, residents do not yet occupy the houses, so they are not yet demanding services. After the residents move in, they routinely demand services in excess of their property taxes, such as roads, schools, and sewer and water infrastructure. This pattern is becoming especially problematic in rural areas as residents increasingly demand services comparable to urban areas.

In contrast, this dynamic does not apply with infill or redevelopment projects because in most cases, the water, sewer, and road infrastructure is already in place, schools are built, and the level of services have been established. In addition, infill or redevelopment projects are typically built at higher densities, which cost less than their lower-density counterparts. For example, the cost of providing services (streets and utilities) to a townhouse at 10 units per acre is less than $10,000 but is more than $32,000 for a house at one unit per acre.[42]

Communities can better reflect the costs of new development and the public infrastructure investment that it requires by requiring new urban fringe development to pay for the full cost of providing services to those areas. Some municipalities that are experiencing rapid growth and development are already assessing full fees to developers to cover projected expenses for roads, schools, sewer, and water infrastructure. In doing so, state and local governments provide an incentive for development in existing communities where infrastructure already exists.

Policy 23. Use compensation fees to address high-priority water quality problems

Some government agencies are using compensation fees when developers or homeowners have difficulty fully meeting a regulatory requirement, such as reducing the quantity or the strength (concentration) of a particular pollutant. Typically, a cost-effective amount of the pollutant is cleaned up, and a fee is assessed for the remaining amount. Then, the state or locality can use funds from the compensation fees to address high-priority water quality issues elsewhere.

For instance, some communities face significant water quality problems in their urban centers. These issues could be related to failing infrastructure, insufficient capacity, point sources, or other past performance problems that cannot be linked to a responsible party. Such water quality issues could prevent a community from redeveloping its brownfield sites, converting surface parking lots to mixed-use developments, or otherwise increasing densities.

Practice Tip: The Maine Department of Environmental Protection established a nonpoint source reduction program to allow an applicant to pay a compensation fee in lieu of meeting certain phosphorus reduction requirements. This program was designed to provide assistance to homeowners and smaller developers who are required to reduce phosphorus loadings from their site. In many cases, the cost of reducing the loads to the required level was not cost-efficient for the amount of phosphorus that would be removed. The compensation fee program permits owners and developers to pay a fee proportional to the level of phosphorus they are unable to remove. The state then can assess where the most urgent phosphorus removal issues are and address those issues using program funds.[43] This program gives the state resources to address the most serious phosphorus problems, which are often found in dense urban centers. As a result, Maine has the tools to direct development to existing communities and mitigate its potential environmental impact.

Policy 24. Charge for water usage on an incremental basis

Research has indicated that residential water users do not pay the entire cost of water and its delivery. In most cases, the local government jurisdictions pay the difference. Therefore, the more water used, the greater the subsidy. Charging for water use on an incremental or block-pricing basis reduces this subsidy.

The latest annual water pricing study conducted by an advisory committee in Fort Worth, Texas, found that residential users were paying nearly eight percent less than the true cost of delivering water.[44] Failure to represent the true cost of delivery is particularly marked in lower-density areas far from central treatment plants, where both water delivery and system expansions are typically subsidized. The cost of delivering water depends both on transmission costs, which increase with distance, and operation and maintenance costs, which increase with the length of systems. Pressure requirements to meet fire codes, for example, are more expensive to maintain across longer, more dispersed networks. As such, average costs often grow as systems cover larger geographic areas, requiring longer system components.

Furthermore, research has demonstrated that more compact communities use less water than lower-density communities, largely as a result of the difference in outdoor water use; homeowners with larger lawns use more water than those with smaller lawns. A 1995 Rutgers University study on New Jersey infrastructure estimated that the cost of providing water to households in conventional dispersed developments was roughly 13 percent higher than the cost of doing so in a more compact context.[45]

Rates that base the per unit cost of water on the consumer's incremental use can encourage conservation and decrease the local government subsidy for lower-density developments. For example, block pricing applies lower per unit costs to base amounts of water use sufficient to meet basic household needs, and incrementally higher rates for additional blocks of water (e.g., the next 5,000 gallons consumed). Such a rate basis would reward homeowners in more compact communities and decrease local government subsidies for water delivery.

Issues to Consider: Decisionmakers must be attentive to the impact of increased water rates by volume on commercial, agricultural, and industrial users and the potential impact that higher rates could have on economic development efforts.

Practice Tip: In North Carolina, a recent drought spurred local water officials in Charlotte to consider whether the imposition of a penalty for excess water consumption would reduce demand. In 2001, after its analysis, Charlotte adopted new fees for residential, multifamily, and commercial water users. Since lawns can be responsible for as much as 60 percent of water usage in some areas, Charlotte's revised pricing system, in effect, lessened the appeal (and value) of a large lawn and landscaping. The county estimates that the average Charlotte household uses 1,100 cubic feet of water per month—approximately 74,800 gallons (equivalent to filling two swimming pools). The new tiered system takes effect once the household use reaches 1,700 cubic feet per month—the rate increases from the base of $1.09 per hundred cubic feet to $1.82 per unit. At 3,200 cubic feet per month, the rate increases again to $3.70 per hundred cubic feet.[46]

Policy 25. Use Clean Water State Revolving Funds for smart growth initiatives

Traditionally, Clean Water State Revolving Funds (SRF) are used to construct and upgrade infrastructure to maintain water quality. As another option, states use SRF funds for other efforts likely to impact water quality, such as comprehensive plans or open space preservation.

Photo courtesy of USDA NRCS.

The SRF is a widely available financing source used to fund municipal wastewater treatment and drinking-water projects, as well as nonpoint source pollution control and estuary protection projects. The states disburse the federal SRF funds to eligible localities and projects in the form of low-interest, long-term loans. Despite the fact that the projects eligible for SRF funds are typically capital expenditures for compliance with national primary drinking-water regulations or projects funding wastewater treatment, the program is flexible enough to allow a portion of the funds to be used for some of the principles of smart growth, such as open space preservation or infill development. For example, the Commonwealth of Massachusetts actively limits the use of SRF funds to support new growth. Collection systems projects are eligible only if 75 percent of the flows existed as of April 1995. Thus, no more than 25 percent of the capacity of a project can be used for new growth.[47]

SRF funds can be used to preserve open space and to create recreational spaces.

Since SRF funding decisions can affect development patterns, states can coordinate their management of SRF loans with emerging smart growth policies and initiatives. States can leverage smart growth benefits out of existing SRF resources by granting additional funds for smart growth enhancements to traditional projects or providing technical assistance on smart growth to project applicants. States could also require long-term comprehensive growth plans, or encourage limits on sewer connections or capacity for new growth in designated areas.[48] Funds also could be used to support and create incentives for comprehensive planning and maintenance of existing water infrastructure.[49]

Issues to Consider: SRF program officers must understand the program and its potential connections to smart growth in order to coordinate the management of SRF funds with broader growth initiatives. SRF program managers might first want to consider whether the use of SRF funding has encouraged growth in areas where growth should instead be discouraged. For example, has SRF funding provided wastewater treatment capacity enabling growth in a source water protection area? Or has SRF-funded wastewater treatment capacity made it more economically attractive for developers to build in areas that might be better left as open or green space? If communities find that the answers to the above questions are cause for concern, SRF managers can be educated to better consider what role their programs could play in supporting smart growth initiatives. At a minimum, the SRF must ensure that projects receiving funding meet the environmental review requirements of the CWA, but it might also hold the potential to achieve other, broader growth objectives at the same time.

WANT MORE INFORMATION?

EPA developed guidelines for using state revolving funds to support smart growth activities. Details can be found at: <www.epa.gov/owmitnet/cwfinance/cwsrf/smartgro.pdf>.

WANT MORE INFORMATION?

For more information on Iowa's Drinking Water SRF program, visit: <www.state.ia.us/epd/wtrsuply/srf/srf.htm>.

Practice Tip: In 2002, Iowa created the Smart SRF for Iowa Clean Water program. This program allows the use of the state's drinking water SRFs for smart growth initiatives, including brownfields cleanup, watershed management, low-impact development practices, and riparian land conservation. The Iowa Finance Authority and the Iowa Department of Natural Resources launched the initiative to change the state's nonpoint source protection plan and the SRF statute to allow the use of SRF funding for smart growth projects. At the time of this publication, the city of Des Moines was exploring the option of using SRF funding for the redevelopment of a 1,200-acre brownfield along the Des Moines River.[50]

Policy 26. Improve oversight of onsite treatment systems

Onsite waste treatment systems (also known as septic systems) are underground tanks that collect, treat, and disperse small volumes of wastewater, traditionally from an individual house. Historically, houses in rural areas distant from sewer collection and treatment systems have been served by septic systems, except in areas with sensitive groundwater or where soils do not allow the treated waste to percolate down.

According to a July 2003 report,[51] decentralized systems are used in 25 percent of all homes in the United States and in 33 percent of new developments. Yet, improperly managed onsite systems can pose environmental challenges. More than half of the existing systems were installed 30 or more years ago, and each year, at least 10 percent of all systems fail. States report failing septic systems as the third most common source of groundwater contamination. Therefore, EPA, states, and localities are increasing efforts to control failure rates through aggressive outreach and, in some cases, permitting programs. The focus in all of these programs is improved and effective maintenance and operation.

WANT MORE INFORMATION?

EPA provides information on creating management districts to oversee onsite systems. Details are available at: <www.epa.gov/owm/mtb/decent/download/guidelines.pdf>.

Decentralized systems can support smart growth in rural areas, or in mountain and coastal areas experiencing growth in the number of second homes. In areas where clustering homes and conservation subdivision design are growth tools, localities are likely to experience better operation and maintenance in onsite systems, as several homes are responsible for and dependent on their functioning. These designs can also conserve open space and reduce the amount of other infrastructure needed to serve new development.

Issues to Consider: Without careful planning, the use of onsite wastewater systems can foster low-density, dispersed development patterns. The decision to install onsite systems must take into account a variety of factors, such as soil conditions, development repercussions, and the likelihood of appropriate maintenance practices. Decentralized systems often occur in rural areas where few development regulations exist. Because of this, local governments might need to increase the type and level of oversight to include permitting, inspections, and operation and maintenance agreements. Otherwise, onsite systems could encourage a lower-density and high land consumptive development pattern.[52]

Policy 27. Provide a stormwater fee credit for redeveloping existing impervious surfaces

Most state and local water quality requirements do not take into consideration the condition of a site before development or redevelopment. By considering pre-development conditions, state and local governments have an opportunity to provide pollution credits or otherwise recognize redevelopment sites as smart choices for preserving water quality. Doing so might provide a greater incentive to redevelop previously developed sites, such as brownfields or greyfields.

In many cases, redeveloping a brownfield or greyfield site will not increase the net contribution to stormwater runoff. A 50-acre parking lot generates the same, if not more, stormwater runoff before it is redeveloped than afterward. For example, Mizner Park in Florida is a former shopping mall that was redeveloped into a mixed-use community. Redesigned from its original pattern of a large retail structure surrounded by surface parking lots, the 29-acre site now includes 272 apartments and townhouses, 103,000 square feet of office space, and 156,000 square feet of retail space. Before redevelopment, the site was 100 percent impervious cover. After redevelopment, impervious cover decreased by 15 percent.

Existing impervious surfaces, such as parking lots, can be transformed into pathways, community gardens, or other neighborhood amenities.

The redeveloped site now includes a long, wide plaza that runs the length of the development and includes grass, trees, and other native landscaping, reducing impervious area. In addition, the developer incorporated numerous small areas for landscaping and trees throughout the site. Stormwater runoff decreased accordingly. Additionally, the redevelopment of brownfield and greyfield properties maximizes return from existing water infrastructure, roads, transit, and other services. Redevelopment of previously developed land also reuses already compacted, disturbed, or impervious soil rather than impacting other soils.

Photo courtesy of USDA NRCS.

Taking pre-development conditions into account, states could develop specific criteria for waiving or reducing current stormwater requirements under certain pre-construction conditions (e.g., redevelopment of an existing surface parking lot). This approach could encourage redevelopment of underutilized properties and maximize the use of existing impervious cover, already degraded soils, and existing infrastructure. Such a waiver could be incorporated into a stormwater ordinance, a state's stormwater management guidance manual, a municipality's public facility manual, or local permitting requirements.

Policy 28. Tie bonds to performance measures

Developers are required to meet certain short-term water quality requirements during and after construction, such as reducing sediments and runoff leaving the site. But there is no mechanism in place for accountability if the developer fails to meet those water quality requirements. Because enforcement of water quality requirements is often carried out by random spot checks, some problems, such as lakes or streams becoming clogged with sediments, are not identified until after construction is completed. As a remedy, communities could require developers to purchase bonds or set aside money to be used to clean up or otherwise comply with water quality requirements, if a regulatory authority discovers within a fixed period of time that those requirements were not met.

States or municipalities could provide developers with incentives to ensure that water quality on their sites is protected through the use of performance bonds. Similar to the approach used for heavily polluting industries, in which businesses are required to purchase surety bonds to cover the costs of future cleanups (should they occur), developers could be required to purchase a bond that is linked to performance measures that monitor water quality impacts on nearby waterways. Under this type of performance bond system, the developer would profit if the water quality is maintained or improved. However, if water quality deteriorates as a result of site-level features, such as large volumes of polluted construction runoff, then the bond money would be spent on cleanup. Given a stake in the future performance of the development's water quality, developers would have more incentive to incorporate cost-effective, long-term water quality protection methods into a project. Such methods could include design elements that rely on natural processes for water quality management, such as buffers or reduced impervious surface areas.

Practice Tip: Officials in Columbus, Ohio, are evaluating the adoption of performance measures for the city's streams and holding area developers responsible for maintaining the streams' water quality. In this scenario, the municipality would create a performance standard for the waterbody. During the permitting process, the developer would be required to put a set amount of money into an account for five to 10 years. This money would be returned to the developer if the stream continues to meet water quality standards at the end of that period.

Policy 29. Use private activity bonds to finance projects that protect water resources

Many local governments issue private activity bonds to private parties in a partnership to finance capital improvements. Such bonds can be a cost-effective way of financing infrastructure projects that protect water resources. Local governments could prioritize projects that receive such financing to encourage projects that will improve existing infrastructure, rather than financing projects that create new infrastructure and growth on the fringe. For example, Florida issues private activity bonds for projects upgrading existing drinking-water and wastewater facilities to encourage additional development where infrastructure already exists.[53]

Issues to Consider: Drinking-water and wastewater facilities generally are exempt facilities under private activity bond regulations and therefore are eligible for tax-exempt status. However, there are federally mandated caps on the amount of tax-exempt private activity bonds that can be issued in a state. States can prioritize the allocation of bonds so that projects that implement smart growth strategies and water infrastructure are more likely to receive bond financing.

Practice Tip: Florida's Growth Policy Act, adopted in 1999, recognizes infill development and redevelopment as important to promoting and sustaining urban cores. Florida's definition of urban infill and redevelopment areas includes those where public services such as water and wastewater, transportation, schools, and recreation are already available or are scheduled to be provided within an adopted five-year schedule of capital improvements. A local government with an adopted urban infill and redevelopment plan may issue revenue bonds and employ tax increment financing for the purpose of financing the implementation of the plan. Areas designated by a local government as urban infill and redevelopment areas are given priority in the allocation of private activity bonds.[54] By giving infill projects priority over other projects (such as greenfield development), the use of existing impervious surface is maximized rather than using bonds to fund development in undeveloped areas.

Policy 30. Allocate a portion of highway and transit funding to meet water quality goals

Water quality conditions are generally not included in transportation funding criteria. Given the numerous connections between transportation-related infrastructure and water, however, states might want to consider water quality criteria when determining funding for proposed transportation-related projects.

The links between transportation, development, and water quality are numerous. Not only do transportation projects influence surrounding development, but the transportation-development nexus also affects runoff pollution in the watershed. Deposition of mobile air emissions into nearby waterbodies is also part of the close relationship between transportation networks, development patterns, and their many impacts on natural resources.

In areas where air quality violates one or more Clean Air Act standards, "conformity" rules require that transportation plans, programs, and projects must not produce new air quality violations, worsen existing violations, or delay timely attainment of Clean Air Act standards. Under conformity, transportation projects cannot be approved, funded, or implemented unless metropolitan planning organizations (MPOs) provide a transportation investment plan that will result in conforming air quality. The MPO's transportation investment plan must conform to its air quality plan, so that when transportation projects are completed, they will not contribute to unacceptable air quality. Similarly, MPOs could include in their analysis of transportation projects a demonstration of how current and projected water quality conditions comply with state and local water quality requirements.

If water quality standards are currently not met and the proposed transportation project would add more pollution to already polluted waters, the MPO could deny transportation funding on that basis. However, the analysis would have to include a comparison of the alternatives in terms of risks to regional water and air quality goals. For example, a proposed transportation project in a highly developed area that supports infill or brownfield redevelopment could reduce total miles driven and subsequently minimize air emissions when compared to alternative development scenarios that have the potential to place the development further out in the metropolitan area and away from transit choices. Or, the same proposed transportation project might increase site-level runoff, but less so than other transportation-development scenarios.

Policy 31. Establish a community preservation fund

Communities might want to consider setting up a fund to specifically target resources to preserve open space, both to improve water quality and to encourage development in an existing community, rather than on its outskirts.

Capital for preserving open space can be generated or set aside by localities through a community preservation fund. Revenue for the fund would come from property taxes and could be matched by a dedicated state fund. By creating such a fund, communities would be taking steps to protect water resources by preserving areas that provide important natural processes, such as filtering pollutants, for maintaining healthy water quality.

Practice Tip: In September 2000, Massachusetts passed the Community Preservation Act, which allows communities to create a local Community Preservation Fund in the municipality funded by a surcharge of up to three percent of the real estate tax levy on real property. Once adopted locally, the act would require at least 10 percent of the money raised to be distributed to three categories: historic preservation, open space protection, and low- and moderate-income housing. The act also annually creates a significant state matching fund of more than $25 million, which will serve as an incentive to communities to take advantage of the provisions of this legislation.[55] As of May 2003, 61 of the 109 communities that held ballot votes passed the act.[56]

WANT MORE INFORMATION?

In 2003, the National Association of Local Government Environmental Professionals, the Trust for Public Land, and Eastern Research Group published *Smart Growth for Clean Water: Helping Communities Address the Water Quality Impacts of Sprawl*, which describes land conservation, watershed management, brownfields redevelopment, and other smart growth tools as key strategies for achieving water quality goals. The document is available at: <www.nalgep.org/publications>.

Policy 32. Establish a clean water management trust fund

Funds for community water management come from many federal, state, and local funding sources. Communities can set up a fund to target resources to manage water runoff and encourage development within the existing community, rather than on the outskirts.

Trust funds can provide additional funding needed to finance smart growth projects that will help protect water resources. Money from a clean water management trust fund, for example, can go towards smart growth development projects such as acquisition of greenways, towards interest on loans for downtown redevelopment projects, or to encourage development on existing impervious surfaces, such as brownfields, rather than developing on green space.[57]

Parkland and natural vegetation buffer an urban stream at Fairview Village in Portland, Oregon.

Photo courtesy of U.S. EPA.

A trust fund can be created by state assemblies, municipalities, nonprofit organizations, or others using revenues from fines, penalties, user fees (e.g., tax on water use), lottery proceeds, taxes on pollution sources, or general assembly appropriations. For example, the Nebraska Environmental Trust Fund receives 49.5 percent of the profits of the Nebraska Lottery after the first $500,000 awarded. These proceeds have annually generated roughly $8.5 million for grant assistance.[58]

Practice Tip: The North Carolina Clean Water Management Trust Fund, created in 1996, provides grants to local governments, state agencies, and conservation nonprofits to help finance projects that specifically address water pollution problems. The fund is supported by appropriations from the General Assembly. At the end of each fiscal year, 6.5 percent of the unreserved credit balance in North Carolina's General Fund (or a minimum of $30 million) goes into the fund. The 18-member independent Board of Trustees has full responsibility for the allocation of resources from the fund and approved more than $31 million in grants in 2003. Grants are provided for projects that enhance or restore degraded waters, protect unpolluted waters, and/or contribute toward a network of riparian buffers and greenways for environmental, educational, and recreational benefits. Projects funded include greenway and open space acquisition, improvements to wastewater treatment facilities, stormwater management, removal of septic tanks, and wetlands and stream restoration.[59]

Policy 33. Offer incentives for adopting land use changes under a TMDL implementation plan

States are required to develop an implementation plan for Total Maximum Daily Loads (TMDLs), but are not required to execute it, as most activities outlined in the implementation plan are completed at the local level. As part of a TMDL implementation plan, states could offer incentives to communities that adopt land use changes that foster smart growth.

Sometimes there are barriers to fully executing the implementation plan at the local level. Obstacles could take the form of industry backlash at the cost of pollutant removal strategies, unexpected increases in pollutant loads due to development, or several years of unusually wet weather, causing unusually high runoff and associated pollutant loads. States can increase the chances of implementation, however, by including a provision in their TMDL requirements that requires full execution of the plan. In addition to this requirement, states can provide guidance and recommendations to communities on how they can support and advance the implementation process. For example, communities that take steps to mitigate the water quality impacts—both at the site and regional level—of their growth decisions would go a long way towards achieving target loadings of some TMDLs. States could detail what land use changes they would like to see implemented, such as more compact site designs, transit-oriented development, larger riparian corridors, or larger areas of open space incorporated into the urban and suburban fabric. To encourage communities to act, states could offer these communities "bonus" points on any applications for CWA Section 319 or SRF funding, or other state-allocated funding sources. Although the bonus points would not guarantee a successful application, they would give an advantage to those communities that implemented the land use mitigation measures over those communities that did not.

ENVIRONMENTAL REGULATORY INNOVATIONS (INCLUDING VOLUNTARY INCENTIVES)

The CWA sets national goals for water quality and process requirements for attaining them. EPA issues federal regulations as part of its role in administering the CWA and delegates specific authority to states and tribes as to how they will attain and enforce federally established standards. For those states and tribes that do not have delegated authority, EPA regions are responsible for establishing, implementing, and enforcing state standards and requirements.

Within this federal-state-local framework for implementing the CWA, there are a number of opportunities to use smart growth approaches to meet state and local water quality goals. The following policies describe opportunities for communities to leverage smart growth approaches to meet current water quality regulations.

Policy 34. Create performance-based standards

Many water quality standards are technology-based. For example, a regulation might call for a detention pond of a particular size, according to the lot size. To provide developers with more flexibility in meeting water quality standards, policymakers may consider the use of performance-based standards that set target goals—such as a 40 percent reduction in stormwater runoff—but leave it to the developer to determine the means by which this goal is achieved. This approach shifts the focus from technologies to the actual reduction of pollutants, and it might encourage implementing land use and zoning policies to achieve water quality goals.

For example, performance-based measures would allow leeway for revised zoning codes or regional plans to redirect development to achieve water quality improvements. These measures might consider the stormwater runoff benefits associated with higher-density development that leads to an overall lower level of imperviousness. Regulations can be supplemented with performance-based standards to provide more flexibility and encourage innovation.

Issues to Consider: Flexible codes might require a significant shift in how government agencies operate. Governments might need to educate the review staff on the principles of adaptive management—identifying and adapting policies based on modeling, monitoring, and other research and analysis efforts. In addition, the adoption of performance-based standards will require the sound use of scientific information to set desired levels of performance and measure the capacity of participants to achieve them.

Practice Tip: Lacey, Washington, passed an ordinance encouraging "zero effect drainage discharge" and "zero effective impervious surface" by revising its building code to specifically encourage and allow development that yields these impacts. It is the first such ordinance in the United States. Under the ordinance, a zero effective impervious surface means "impervious surface reduction to a small fraction of that resulting from traditional site development techniques, such that usual manmade drainage collection systems are not necessary."[60] The ordinance allows prescribed stormwater control requirements to be waived when project design uses alternative techniques to reduce stormwater runoff. Possible design approaches allowed under the ordinance include: replacing all driveway and parking areas with pervious materials, planting native landscaping with greater capacity to slow runoff and take up the water, allowing for smaller rooftop exposures and/or rooftop gardens, or constructing narrow roadways with substantial vegetative berms.[61]

Policy 35. Consider future growth when developing TMDLs

States are responsible for establishing water quality standards for their waterbodies, including TMDLs for pollutants when a waterbody or water segment is impaired. This process might consist of guidance for local governments on how to comply with the federal TMDL requirements, or the development of new state standards for developing and/or reviewing TMDLs to ensure that the regulations are followed. Often, future growth is not considered or specified in state guidance on TMDLs.

Allocating impacts from future growth is currently not required at the federal level; however, some states require the inclusion of future growth in TMDL calculations. As such, their guidance documents represent an opportunity to include current and future land use decisions within the TMDL process. States may include additional TMDL component requirements that would ultimately help them achieve the final target loading. In this context, for example, states could require that the development-related impacts from future growth be considered when developing TMDL allocations. The inclusion of future growth would help states meet their TMDL targets and favor less-polluting smart growth development options.

Practice Tip: Georgia, as part of its TMDL process, requires any locality asking the state for an environmental permit that facilitates growth and development (e.g., wastewater or water withdrawal permit) to conduct a watershed assessment. These assessments provide additional information on point and nonpoint pollution sources. Applicants must identify pollution sources, model future land use scenarios, and provide solutions to water quality problems.[62]

Policy 36. Make adequate water a prerequisite of additional growth

Local permitting and approval processes for development often do not explicitly consider available water supplies when evaluating potential development. Incorporating provisions to do so can help communities ensure that future development will not overburden existing water resources.

Local decisionmakers may want to assess potential impacts on future water supplies and quality prior to permitting new developments. Such assessments could provide early warnings if a new development will likely have an unacceptable impact on water quality and water supply. These assessments will be most effective if completed early in the planning process, by connecting water supply plans to comprehensive plans, as well as at the point of permitting, when the impact of a specific proposed development can be estimated. By making such analysis a routine part of planning for large-scale growth, decisionmakers can help ensure that future water supplies will be adequate, and that water quality will not be compromised by growth.

Moving in this policy direction, the Charles River Watershed Association completed an environmental assessment for a zoning plan in the town of Holliston, Massachusetts, that could link future growth to sustainable water supplies. The assessment used geographic information systems (GIS) to map areas of developable land that are critical for replenishing aquifers. The association calculated a "water budget" for the town, showing the impact of various levels of development on water resources. Such planning can prevent future water supply shortages and ensure that new developments have the necessary water infrastructure.[63]

Practice Tip: A new California state law, effective January 2002, requires all developers of proposed projects of 500 or more homes to demonstrate that ample water supplies exist prior to construction.[64] Cities and counties are prohibited from issuing permits for the construction of projects unless the local water agency verifies that it has enough water to serve the new growth at least during the next 20 years.[65] This process allows water suppliers to refuse to serve additional houses to prevent shortages that could affect existing customers. In some cases, it could require developers to help find and pay for new water sources. Although the bill does not directly encourage the use of compact development, it does so indirectly because more compact development usually consumes less water on a per household basis. In addition, the bill offers a waiver for projects in infill areas, where projects are most likely to incorporate compact building techniques. As a result, the new law has the potential to indirectly reduce household water demand and site runoff. Additional provisions might be necessary, however, to ensure that developers do not evade the law by proposing 499-unit projects when water supply is in doubt.

Policy 37. Incorporate smart growth into stormwater management plans

Communities are mandated to develop stormwater management programs under the National Pollution Discharge Elimination System (NPDES) requirement. Some components of a regional smart growth program can be used to meet or enhance a community's requirements for a stormwater management program.[66] For communities that have already adopted smart growth plans, recognizing the water benefits of those plans and making them part of the stormwater water plan submission can be a low-cost way to meet some of the stormwater management program requirements. In addition, communities that have not yet adopted smart growth plans might want to investigate smart growth approaches that can help them meet stormwater management program responsibilities and meet other community goals with the same investment.

Photo courtesy of USDA NRCS.

In response to the 1987 amendments to the CWA, EPA developed Phase I of the NPDES Stormwater Program in 1990. The Phase I program addressed sources of stormwater runoff that had the greatest potential negative impacts on water quality. Under Phase I, EPA required NPDES permit coverage for stormwater discharges from medium and large municipal separate storm sewer systems (MS4s) located in incorporated places or counties with populations of 100,000 or more, and for construction sites that disturb 5 or more acres. The Phase II Final Rule requires NPDES permit coverage for stormwater discharges from small municipal separate storm sewer systems and for construction sites that disturb between 1 and 5 acres. A stormwater management program requires six minimum control measures (MCMs), including:

Public education and outreach are required under Phase II and can help support a community's revitalization goals.

1. Public education and outreach

2. Public participation/involvement

3. Illicit discharge detection and elimination

4. Construction site runoff control

5. Post construction runoff control

6. Pollution prevention/good housekeeping

A community's smart growth plan can help fulfill many of these minimum control measures. For example, the Washington State Department of Ecology developed a model permit for communities that must comply with EPA's Stormwater Phase II regulations. The permit lists infill development policies as a creditable policy to mitigate post-construction stormwater volumes.[67] In addition, an effective smart growth planning process will necessarily involve public outreach and involvement on future growth areas, and that discussion should involve water quality impacts. Thus, smart growth planning helps fulfill MCMs One and Two. As discussed in the sections above, smart growth effectively reduces development footprints for a given amount of development, reducing runoff both during and after construction, further fulfilling MCMs Four and Five.

Practice Tip: Jackson County, Michigan, has been able to take advantage of the smart growth and Phase II interactions. In 2003, local officials created the Upper Grand River Watershed Initiative. Even though the initiative was created to address Phase II requirements, the plan architects recognized the smart growth benefits of this plan. For example, although education and public awareness are a large part of the plan, it will likely touch on issues such as land use, urban sprawl, brownfield, redevelopment, wetlands preservation, and zoning regulations.[68]

Policy 38. Incorporate smart growth into pollution trading programs

Trading allows a community to use a market-based approach to maintain its water quality. Trading is based on the idea that different sources face different costs to control the same amount of a given pollutant. Trading therefore allows the sources, such as facilities or nonpoint sources, facing higher pollution control costs to meet their required reductions by purchasing equal (or better) reductions from another source.[69] Trading then achieves the same water quality improvements at an overall lower cost. Trading might also benefit impaired urban waterways where reaching healthy levels is difficult. In 2003, EPA announced a new Water Quality Trading Policy, which is designed to further reduce industrial, municipal, and agricultural discharges into waterways.[70] The policy provides guidance to states and tribes on how trading can occur under the CWA and its implementing regulations.

Numerous opportunities to incorporate smart growth approaches into a trading framework exist. At the state level, for example, where trading policies are determined, states can consider calculating pollutant loads on a per housing unit basis rather than the more conventional per acre basis. By being able to calculate loads on a housing-unit basis instead of on an acre basis, communities are better able to account for the water quality benefits of higher-density developments. In addition, the current trading policy allows states to consider disturbed land in addition to or instead of overall percent impervious cover. Under this option, communities will be able to give credit to developers who use a compact site design and disturb less land than a typical low-density development.

Given the potential water quality benefits of better site design, this trading policy could provide some communities with relief. For example, if a community has several sources for which it is costly to further reduce loadings, the point source can "buy" credits from a developer who is considering a compact site design. The funds provided by the point source can then be used for design assistance to further enhance those site design practices that achieve smart growth. This type of relationship might provide incentives for additional developers to implement better site design practices when they realize the water quality "savings" are marketable.

WANT MORE INFORMATION?

The World Resources Institute developed a trading Web site to provide a simple way for buyers and sellers to connect. It is located at: <www.nutrientnet.org>.

Practice Tip: The Cherry Creek Reservoir Watershed Phosphorus Trading Program in Denver, Colorado, is an example of an innovative point/nonpoint source trading program.[71] The goal of this program is to allow point source discharges to increase within a TMDL cap. To help reach this goal, point and nonpoint controls have been implemented to reduce phosphorus loadings in the watershed. Municipal facilities must now optimize controls, comply with permit limits, and implement best management practices (BMPs) for urban runoff before a trade is approved.[72] Credits generated from nonpoint source pollutant reduction facilities can be used to offset growth when a need is demonstrated. Development and credit use must be consistent with a basin plan established by the Cherry Creek Basin Water Quality Authority under legislative mandate. Furthermore, permits issued by the state must also be consistent with the basin plan and use of credits approved by the Authority.

Policy 39. Use smart growth to vigorously pursue CWA antidegradation policy

The CWA requires states to have antidegradation policies and implementation methods in place to maintain the health of waterbodies. Antidegradation is part of a larger process of protecting waterbodies that involves setting water quality standards. States or EPA must first designate uses for targeted waterbodies, then develop water quality criteria to protect those uses, and finally place the better quality streams into higher antidegradation tiers: Tier II for high-quality waterbodies and Tier III for exceptional value or outstanding waterbodies. Discharges into these waterbodies will be more tightly controlled, wetlands and natural habitats will be preserved, and stormwater will be recharged into the ground instead of eroding stream banks. Development can occur, but only if the quality of waterbodies and wetlands are maintained.

Although antidegradation goals and requirements are clearly stated in the CWA, many states and communities are still formulating their specific responses to antidegradation. A smart growth approach can facilitate compliance in several ways. First, by accommodating the same amount of growth on less land than conventional low-density development, smart growth allows certain areas of a watershed, which might otherwise be developed, to be set aside to preserve existing water quality. Second, where portions of a watershed will be developed, antidegradation policy requires cost-effective controls, and smart growth offers a highly cost-effective approach to minimizing the amount of degradation. Smart growth reduces the cost of

Antidegration measures can help preserve pristine waters.

Photo courtesy of USDA NRCS.

infrastructure (e.g., roads, water, sewer) compared to conventional lower-density development, and can also substantially reduce nonpoint runoff from a given amount of development. Finally, antidegradation policy allows a certain amount of degradation, if necessary, for "economic development." In spite of the fact that most residential development fails to support itself from a tax-revenue perspective, smart growth developments can substantially lower municipal cost burdens, making it an economical way to grow and still comply with the policy.

Policy 40. Create a sliding scale of mitigation requirements based on level of density

Stormwater regulations typically do not recognize the benefits that can result from denser developments, particularly those in existing communities. Required runoff reduction is traditionally based on acreage and applied to all development projects—regardless of location within the region or the density of the development. Thus a 5-acre, high-density redevelopment of a parking lot accommodating 100 units is often required to reduce the same amount of runoff as a 5-acre, low-density development accommodating five units. Instead, communities can implement a sliding scale for stormwater mitigation based on the development's density level. This approach will recognize the stormwater benefits that can result from more compact developments.

More compact, mixed-use developments generally require less land and cause fewer water quality impacts than their conventional, less dense counterparts. When compact developments are located in existing communities—thereby reducing the pressure for development of sensitive ecological areas such as headwaters, wetlands, riparian corridors, and floodplains—their stormwater benefits are greater still. As a result, these compact developments in existing communities reduce the need for stormwater mitigation that otherwise would have been required with conventional developments.

Communities can encourage compact development by reducing mitigation require-ments based on density. This approach provides a financial incentive for higher-density (more compact) development that will further reduce a community's overall needs for stormwater mitigation. For example, a state or municipality can set a pol-lutant reduction target for new development that incorporates a sliding scale accord-ing to the project's density (see Figure 2). The higher the density, the less stringent pollutant reduction requirements would be. Residential, commercial, business, or mixed-use redevelopment at any density could be credited the full amount of pollutant removal, thereby waiving responsibility for any additional mitigation efforts as a result of new development. As the requirements for removal efficiency increase with lower-density projects, so do the costs of mitigation, thus providing financial incentive for higher-density projects.

Figure 2: Example of Possible Land Use Water Quality Credits

Land Use Density (housing units/acre)	BMP Removal Credit (%)
Redevelopment (post-redevelopment imperviousness = current imperviousness) at any density	50 to 75
Single-family residential (1 to 5 units per acre)	0
Residential (5 to 10 units per acre)	15 to 20
Medium-density residential (11 to 25 units per acre)	25 to 35
High-density residential (> 25 units per acre)	35 to 50

Policy 41. Modify facility planning area process to support smart growth

Facility planning areas (FPAs), authorized by the CWA, call for states to integrate and coordinate planning for wastewater systems to better protect water quality. The provision seeks to manage the placement and timing of wastewater system expansion or construction, and evaluate any potential environmental impacts. When planning for wastewater services, the FPA provision requires water and sewer providers to forecast future population growth (and therefore development and water infrastructure), due to its direct and significant impact on the community's capacity to manage land use planning to reduce water demand. In addition, a state or its designated agent can deny wastewater system expansions through the FPA approval process, giving states a clear role in managing growth.

The FPA process also highlights an important role for states in managing water. Historically, the primary environmental concern of facility planning was the effect of non-regulated wastewater systems on water quality. Now a larger environmental concern is the effect of the rapid dispersion of people and jobs to outlying areas. The conversion of land from open space to development creates nonpoint source pollution and endangers water resources once thought secure from pollution threats. [73]

Facility planning area processes can better account for these impacts through explicit provisions that support the expansion of wastewater systems—and therefore future growth—in existing communities or those characterized by compact development. Communities (usually municipalities or sanitary districts) are required to identify geographic areas currently served by wastewater systems, as well as those in need of service within the next 20 years. States can, in addition, require these FPAs (also known as sewer service areas or sewerage service agencies) to comply with local plans that encourage reinvestment in existing areas. In addition, FPAs must apply to the state or their designated agent for approval of amendments to or expansions of their existing service areas.

States could further support communities by using evaluation standards that favor plans to expand sewer service to areas slated for compact, rather than dispersed, development. All these policy innovations build on current requirements for states to consider the environmental impacts of wastewater system expansions. The innovations recognize the potential water benefits (both in terms of quantity demanded and system efficiency) associated with more compact growth. In so doing, they serve as an opportunity for states to fulfill their water management duties and simultaneously support communities' attempts to achieve smart growth.

Issues to Consider: Illinois' recent experience highlights some of the more difficult issues that can arise from the FPA process and some dramatic changes that might result. The FPA process in Illinois had, over time, resulted in a great deal of frustration—some municipalities considered the state role in their growth planning to be inappropriate; developers claimed that the process added time and expense to their efforts; and some "no growth" advocates claimed that the FPA process did too little to manage growth, particularly in sensitive environmental areas. In 1998, the Illinois Environmental Protection Agency (IEPA) announced plans to discontinue the FPA process, citing, among other reasons, a growing incidence of inter-jurisdictional battles that it was forced to mediate. In effect, the environmental agency had become an arbiter of community boundaries. [74]

Subsequently, after receiving feedback from a range of parties, IEPA reversed course in 1999, announcing plans to retain the FPA process. This decision was based in part on a comprehensive evaluation of the FPA program by the Openlands Project. This evaluation concluded that, although the program had many flaws, it should be retained and improved. IEPA responded in a September 2002 report, concluding again that the FPA process should be eliminated. Among the principal problems cited by IEPA was the fact that the agency's mediation role (between communities over boundaries and borders) was beyond the purview of the agency. The following month, the IEPA director dismissed the conclusions as final policy, instead inviting public comment on how to resolve ongoing problems with the state FPA process so that it could be retained and improved.[75] By September 2003, IEPA had begun efforts to phase in "a watershed-based approach that will ultimately phase out reliance on the more narrowly focused Facility Planning."[76] Undoubtedly, the state's plans for managing water through coordinated wastewater planning efforts will continue to evolve, providing a lesson for those who would attempt to create an FPA process without many of the difficulties of the Illinois experience.

Practice Tip: The Northeastern Ohio Area Coordinating Agency (NOACA) offers a unique perspective on the potential to link growth and water planning through the FPA process. The agency is designated by the state as the entity responsible for area-wide planning under Section 208 of the CWA—the same act that established the FPA process. In addition to the requirements for wastewater treatment issues, NOACA also considers the nonpoint source pollution impacts associated with growth. Its Clean Water 2000 report establishes the basis for evaluating sewer plans and is guided by principles that seek to "optimize use of existing investment in infrastructure, not encourage public investments in new infrastructure."[77] Such objectives support the goals of the NOACA board to "encourage efficient, compact land use development that facilitates mobility, saves infrastructure costs, preserves environmentally sensitive and agricultural lands, and enhances the economic viability of existing communities within the region."[78]

Finally, the unique dual role of NOACA as the area's metropolitan planning organization charged with the distribution of and planning for transportation resources demonstrates an even more critical connection—the opportunity to connect wastewater and transportation infrastructure planning. Together, both issues exert tremendous influence on how a community grows. NOACA seeks to integrate the two efforts through seven planning strategies that include technical information sharing, the development of models that would generate results useful for both efforts, and uniform standards for use in county comprehensive planning.[79]

EDUCATION

The impact of development on water resources is so vast that regulations alone are insufficient to improve the quality of a community's water resources, hence the importance of education. Local officials, residents, business owners, developers, and other stakeholders might need education on the many ways their actions affect the community's water resources. In particular, given the vast body of federal and state legislative action on water, education can also help stakeholders better understand the goals and objectives of environmental agencies, and the ways in which they can provide assistance to localities and residents.

Opportunities abound for states and localities to incorporate smart growth principles into their program implementation efforts—where education plays an important role. Incentives, best practices, and other approaches to encourage growth in existing communities all work best if communities are educated about needs and goals so that overall runoff is minimized and high-value ecological lands are preserved.

Policy 42. Create partnerships to improve water quality

Municipal responsibility for water resources is often spread throughout several different agencies and departments. In addition to government agencies, the public, developers, construction companies, and others also affect water resources. Partnerships are therefore crucial to ensure a comprehensive and effective approach to smart growth and maintaining water quality. Partnerships can leverage funding, coordinate planning across a region, and share knowledge to better protect water resources.

Development decisions are enhanced when localities engage residents and other stakeholders on how to accommodate growth while still protecting the community's valued water resources. Partnerships between nonprofit organizations, such as land trusts, and governments can be effective in identifying, prioritizing, and eventually acquiring critical parcels for preservation that are under threat of development within watersheds. Educational partners, such as universities and research institutions, can be involved in the development of technology to estimate the potential impacts of development on sensitive water resources. Other partnerships, such as those with foundations or state or federal environmental agencies, can yield important new sources of funding, technology, or technical assistance for localities. Partnerships and ad hoc affiliations of affected groups not only coalesce ideas and energy for water preservation, they also serve to educate all

Partnerships, including schools, can help advance a community's smart growth and water quality goals.

Photo courtesy of USDA NRCS.

members on the many ways in which water resources can be used, abused, and eventually protected.

Issues to Consider: Assembling and maintaining an interagency team and including outside stakeholders can be challenging and time-consuming. Without a clear source of funding, resource considerations can make it difficult for an ad hoc group, for example, to organize and distribute necessary work among members to achieve its objectives. Partnerships of volunteer members or agencies can succeed, however, if efforts are focused on coordinating and achieving discrete, well-defined tasks; enabling each group to contribute in ways related to its strengths; and educating member organizations on the priorities and resources that others bring to the table.

> **Practice Tip:** Rapid, dispersed, low-density development in north central Texas prompted various federal agencies to form the Interagency Stream Team to help communities and developers understand the effects of rapid growth and development on open space, habitat, and streams. Comprised of volunteer engineers, city planners, hydrologists, and other specialists from agencies such as EPA, U.S. Army Corps of Engineers, North Central Texas Council of Governments, Texas Parks and Wildlife, and FEMA, the team provides advice on environmentally friendly ways to manage and restore streams and riparian corridors. The team's project reviews and field visits provide expertise and recommendations concerning project design to municipalities and developers to protect open space, water quality, and habitat. The partnership has provided significant technical support and advice on development, and its guidance and recommended policies are raising the general awareness of maintaining safe and sound aquatic ecosystems throughout the region.[80]

Policy 43. Educate local officials on the water quality impacts from development

Local officials exert a powerful influence over land use development decisions, but might not fully understand the impacts of their decisions on water quality. A decision, for example, to site a new office park on developed land at the urban fringe might seem attractive for fiscal reasons. However, to be able to analyze all aspects of the project, officials need to consider the total cost of expanding water and sewer lines to the new development, the impacts of potential stormwater runoff from the site's large surface parking lots, and the deposition of emissions from commuting office workers into nearby waterways. Training local officials responsible for development decisions, as well as water quality staff, on smart growth and its water quality benefits can help encourage collaboration, resulting in the use of practices and policies that better support mutually shared goals for growth and water protection.

Photo courtesy of the NEMO program and the University of Connecticut.

A town meeting in East Haddam, Connecticut, develops strategies to address local water quality issues.

Given the many aspects of growth that elected officials must consider—such as economic impact, job creation, physical design, and cultural and historical resources—some water quality educators approach water quality through a broader framework of community assets. The experience of Nonpoint Education for Municipal Officials (NEMO), suggests that the concepts of smart growth and community character are often more appealing and tangible to communities than are the water quality aspects of development. Administered by the University of Connecticut, NEMO is a network of local leaders that provides training in watershed management and land use planning to local officials throughout the country.[81] NEMO's approach to education highlights the numerous benefits—including water quality—that smart growth development has to offer.

WANT MORE INFORMATION?

NEMO maintains a Web site that contains fact sheets, reports, presentations, and additional information to educate local officials on the water quality impacts from different land uses. These resources can be accessed at: <www.nemo.uconn.edu>.

Practice Tip: The National Center for Smart Growth Research and Education at the University of Maryland runs a program providing smart growth information to federal, state, and local officials, as well as nonprofit and private firms. The Maryland Smart Growth Leadership program focuses on community development, environmental systems and management, leadership principles, and infrastructure planning, as well as social, economic, and environmental effects.[82]

Policy 44. Develop a model town to demonstrate how and where polluted runoff flows

Many local government officials, planners, and residents are not fully aware of how development contributes to water quality problems. Theoretical understanding of nonpoint source pollution is as important as understanding how the flows of specific local and regional waterways will be impacted by current and proposed developments. One simple way to demonstrate pollution flows from development is to develop a model town.

For example, in Northglenn, Colorado, the local government built a model of the town, which they dotted with food coloring. Water was then sprayed on the model to show how the unfiltered pollutants, as shown with the food coloring, washed over the landscape and through the drainage system into the local stream. The model continues to help educate stakeholders on how different development scenarios impact the environment. It also provides an opportunity to discuss the details and implications of better development models, such as improving housing layouts and designs, creating more compact communities, reducing the footprint of parking lots, and planning for open space for stormwater benefits. Finally, by using small sponges, the model can show residents in older, built-out neighborhoods how they can mitigate the impacts of stormwater on their property. The sponges act as lawns and gardens and are used to show how directing rainwater from the rooftop onto these areas, instead of into the street, decreases pollutants and water that flow into nearby streams.

Policy 45. Create a program to certify developers, builders, and other industry professionals responsible for implementing BMPs

Best management practices (BMPs) provide useful examples to developers, residents, and other stakeholders on how to improve water quality, but they are only as effective as the quality of their implementation. For example, despite the fact that a biorention or a grass swale serves as a BMP for reducing stormwater runoff, its success can be hampered if it is not well-placed (i.e., if it is located over soils that do not percolate well), not well-designed (i.e., it fails to catch significant site runoff), or not well-maintained (i.e., if trash is allowed to collect and accumulate).

One way to ensure that BMPs are effective is to certify contractors who have demonstrated a capacity to construct, implement, and/or manage them well. Such a program provides potential BMP users greater assurance that the maximum benefits will be achieved. It also serves to build a demand for the skills required to attain certification among contractors. Local or state agencies could administer the certification program and provide subsequent random inspections to ensure that the contractors' work is yielding the water quality benefits expected from a BMP.

> **Practice Tip:** Construction activities are known to produce significant nonpoint source pollution as a result of site erosion and runoff. The state of Maine has taken steps to mitigate these impacts by certifying developers who successfully demonstrate the use of techniques for erosion control. Under its Erosion Control Law, Maine's Department of Environmental Protection (DEP) offers this voluntary, incentive-driven certification program to broaden the use of effective erosion control techniques. Contractors are first taught erosion and sedimentation control practices; then, one of their construction sites is inspected to demonstrate their hands-on understanding of erosion control principles. Once this activity is completed, the contractor is certified. As an incentive, the certification program provides free marketing for developers and permits a certified contractor to advertise as a "DEP Certified Contractor."[83]

Policy 46. Provide municipalities with sufficient data to make better land use decisions

> **WANT MORE INFORMATION?**
>
> EPA has a variety of water quality information available at: <www.epa.gov/waters>.

Land use data—such as data from remote sensing or mapping technologies—might not be easily accessible to localities making decisions on where to direct development. Increased coordination of the use and sharing of information, technology, and models between localities and sources collecting the data can help communities make more informed land use decisions.

Environmental agencies, research institutions, and federal agencies collect and analyze a great deal of data and information, but do not necessarily make it easily accessible to localities. Providing municipalities with this information—and the technical capacity to use it to its full benefit—can help local officials and residents make decisions about the long-term impacts of the development decisions they make today. Tools such as GIS and remote sensing are particularly important in the early stages of the planning process (e.g., creating or revising comprehensive plans) when the foundation is laid for growth that will occur during the next several years. For example, remote sensing data that show the growth in sediment throughout time at the base of a river or in a lake as a result of upstream erosion caused by development of previously forested lands can be an incentive to better direct future growth to mitigate impacts. Also, GIS maps can succinctly illustrate the nexus between critical environmental resources and encroaching development pressures, thereby highlighting areas in need of protection.

Practice Tip: Maine's Beginning with Habitat program is a habitat-based landscape approach to assessing wildlife and plant conservation needs and opportunities. The goal of the program is to maintain sufficient habitat to support all native plant and animal species currently breeding in Maine. It accomplishes this by providing GIS data to municipalities. These maps can then be overlaid on town maps to highlight areas where protection efforts should be focused. The maps provide communities with information to guide conservation of valuable habitats and thereby protect water resources.[84]

WANT MORE INFORMATION?

EPA Region 5 and Purdue University developed an online tool, the Long-Term Hydrologic Impact Assessment model, to help planners measure the water quality impacts associated with land use changes. The model is located at: <danpatch.ecn.purdue.edu/~sprawl/LTHIA7/Index.html.>

[3]U.S. Department of Agriculture, Economic Research Service, Natural Resources and Environment Division. 1997. *National Resources Inventory*.

[4]U.S. Department of Agriculture, Economic Research Service. "Major land use changes in the contiguous 48 states." *Agricultural Resources and Environmental Indicators (AREI), 1996-97*. Agriculture Handbook No.712, July 1997.

[5]The 10 percent figure is not an absolute threshold. Recent studies have indicated that in some watersheds, serious degradation may begin well below 10 percent. However, the level at which watershed degradation begins is not the focus of this paper. For purposes of this analysis, the 10 percent threshold will be used as an average figure.

[6]See, for example, Montgomery County Department of Environmental Protection. 2000. *Stream Conditions Cumulative Impact Models for the Potomac Subregion*; Caraco, Deb. 1998. *Rapid Watershed Planning Handbook- A Comprehensive Guide for Managing Urban Watersheds*. Ellicott City, MD: Center for Watershed Protection; Schueler, Tom. 1994. "The Importance of Imperviousness." *Watershed Protection Techniques*. 1.3: 100-111. Ellicott City, MD: The Center for Watershed Protection. <www.stormwatercenter.net/Practice/1-Importance%20of%20Imperviousness.pdf>. Arnold, C.L. and C.J. Gibbons. 1996. "Impervious Surface Coverage: The Emergence of a Key Environmental Indicator." *Journal of the American Planning Association*. 62.2: 243-258.

[7]U.S. EPA. June 2003. EPA's *Draft Report on the Environment Technical Document*. EPA 600-R-03-050. <www.epa.gov/indicators>.

[8]New Jersey Water Supply Authority. Raritan Basin Project Description. <www.raritanbasin.org/Description.htm>.

[9]Denver Regional Council of Governments. "Local governments united on guiding growth: Mile High Compact signed." *Regional Report*. <www.drcog.org/pub_news/releases/RR%20Mile%20High%20Compact.pdf>.

[10]U.S. EPA, Office of Water. Watershed Academy Web site. <www.epa.gov/owow/watershed/wacademy/acad2000/protection/r3.html>.

[11]U.S. EPA, Region 1. Region 1 Smart Growth Web site. <www.epa.gov/region01/ra/sprawl/grants1999.html#10>.

[12]Henry, Natalie. June 13, 2003. "SoCal Counties Combining Habitat Conservation Plans with Clean Water Permits." *Land Letter*. E&E Publishing, LLC. <www.eenews.net/Landletter.htm>.

[13]U.S. Census Bureau, Population Division, Population Projections Program. 2000. *Annual Projections of the Total Resident Population as of July 1: Middle, Lowest, Highest, and Zero International Migration Series, 1999 to 2100*. Washington, D.C. <www.census.gov/population/www/projections/natsum-T1.html>.

[14]Rutgers University. 2000. *The Costs and Benefits of Alternative Growth Patterns: The Impact Assessment of the New Jersey State Plan*. Center for Urban Policy and Research. New Brunswick, NJ.

[15]*Ibid*.

[16]Pollard, Trip. 2001. "Greening the American Dream?" *Planning*. 67 (10): 110-116.

[17]Trust for Public Land. 1997. *Protecting the Source: Land Conservation and the Future of America's Drinking Water*. San Francisco, CA: Trust for Public Land.

[18]Hillsborough County, FL. Hillsborough Greenways Program. <www.hillsboroughcounty.org/pgm/community/greenways_program.html>.

[19]Maryland-National Capital Park & Planning Commission, Montgomery County Department of Park & Planning. Community Based Planning. <www.mc-mncppc.org/community>.

[20]Openlands Project. October 2001. *Protecting Illinois' Environment through a Stronger Facility Planning Process*. <www.openlands.org/reports/FPA%20Report.pdf>.

[21]Wisconsin Department of Natural Resources. Wisconsin's Sewer Service Area Planning Program. <www.dnr.state.wi.us/org/water/wm/glwsp/ssaplan>.

[22]City of Columbus Division of Sewerage and Drainage. November 3, 2000. Facilities Plan Update. <ci.columbus.oh.us/plan>.

[23]North Carolina Floodplain Mapping Program. <www.ncfloodmaps.com/pubdocs/NCFPMPHndOut.htm>.

[24]McElfish, James M. Jr. and Susan Casey-Lefkowitz. 2001. *Smart Growth and the Clean Water Act*. Northeast-Midwest Institute. <www.nemw.org/SGCleanWater.pdf>.

[25]Temple University Center for Public Policy and Eastern Pennsylvania Organizing Project. 2001. *Blight Free Philadelphia: A Public-Private Strategy to Create and Enhance Neighborhood Value*. <www.temple.edu/CPP/content/reports/BlightFreePhiladelphia.pdf>.

[26]Pryne, Eric. May 20, 2002. "20 Years' Worth of County Land?" *Seattle Times.*

[27]Ibid.

[28]Deason, Jonathan, et al. September 2001. *Public Policies and Private Decisions Affecting the Redevelopment of Brownfields: An Analysis of Critical Factors, Relative Weights and Area Differentials.* Prepared for U.S. EPA, Office of Solid Waste and Emergency Response. Washington, D.C.: The George Washington University <www.gwu.edu/~eem/Brownfields/project_report/report.htm>.

[29]Fairfax County Public Works and Environmental Services. 2001. *2001 Public Facilities Manual.* Fairfax County, VA.

[30]Ebert, Karl. August 2002. "New stormwater rules eyed: Proposal calls for easing of regulations in city redevelopment areas." *Oshkosh Northwestern.* Oshkosh, WI: Northwestern.

[31]U.S. EPA. Development, Community, and Environment Division. June 2003. *Minimizing the Impacts of Development on Water Quality* [Draft].

[32]Cooper, Carry. Projects - Mizner Park. <www.coopercarry.com/4/4e1c.html>. Also, City of Boca Raton, FL. Downtown Redevelopment. <www.ci.boca-raton.fl.us/econ/downtown.cfm>.

[33]Metropolitan Washington Council of Governments, Metropolitan Washington Air Quality Committee. August 13, 2003. *Plan to Improve Air Quality in the Washington, DC-MD-VA Region.* <www.mwcog.org/environment/air>.

[34]Alliance for the Chesapeake Bay. 1997. "Air Pollution in the Chesapeake Bay." Baltimore, MD.

[35]Kaspersen, Janice. November/December 2000. "The Stormwater Utility: Will it Work in Your Community?" *Stormwater, The Journal for Surface Water Quality Professionals.*

[36]Cherry Creek Basin Water Quality Authority. Home page. <www.cherrycreekbasin.org>. Also, Cherry Creek Stewardship Partners Web site. Cherry Creek Stewardship Partners. Home Page. <cherry-creek.org>.

[37]Vellinga, Mary Lynne. January 31, 2002. "Sewer Fee Plan to Limit Sprawl Gains Approval." *Sacramento Bee.* <www.sacbee.com/content/news/story/1555934p-1632412c.html>. Also, Sacramento Regional County Sanitation District. Rates and Fees. <www.srcsd.com/costs.html>.

[38]State of New Jersey, Smart Growth Infrastructure Tax Credits Web site. 2002. <www.state.nj.us/budget02/smarttax.html>.

[39]Vermont Agency of Natural Resources, Department of Environmental Conservation, Water Quality Division. February 2001. *Management of Stormwater Runoff in Vermont: Program and Policy Options.* Prepared for the Vermont General Assembly. <www.anr.state.vt.us/dec/waterq/stormwaterWIP.htm>.

[40]ICMA. 1998. *Why Smart Growth: A Primer.*

[41]U.S. EPA. *Our Built and Natural Environments.* EPA 231-R-01-002. January 2001.

[42]ICMA. 1998. *Why Smart Growth: A Primer.*

[43]Maine State Legislature, Office of the Revisor of Statutes. 2003. Title 38: Waters and Navigation. Chapter 3: Protection and Improvement of Waters. <janus.state.me.us/legis/statutes/38/title38sec420-D.html>.

[44]Fort Worth Water Department. September 2002. *Recommendation of the 2002 Informal Water and Wastewater Retail Rate Advisory Committee.* <www.fortworthgov.org/water/Reports/2002ratereportEnglish.pdf>.

[45]Burchell, R.W. and D. Listokin. 1995. *Land, Infrastructure, Housing Costs and Fiscal Impacts Associated with Growth: The Literature on the Impacts of Sprawl Versus Managed Growth.* New Brunswick, NJ: Rutgers University, Center for Urban Policy Research. As summarized in *The Technological Reshaping of Metropolitan America*, Office of Technology Assessment. OTA-ETI-643.

[46]City of Charlotte and Mecklenburg County Government. Charlotte-Mecklenburg Utilities. < www.charmeck.org/Departments/Utilities/Home.htm>.

[47]U.S. EPA, Office of Wastewater Management. October 2000. *Potential Roles for Clean Water State Revolving Fund Programs in Smart Growth Initiatives.* EPA 832-R-00-010. <www.epa.gov/owm/cwfinance/cwsrf/smartgro.pdf>.

[48]Ibid.

[49]U.S. EPA. Development, Community, and Environment Division. Table of Contents, Introduction. *Smart Growth Projects by Statutory Program.* <www.epa.gov/smartgrowth/pdf/sg_at_work.pdf>.

[50]National Association of Local Government Environmental Professionals, The Trust for Public Land, and Eastern Research Group. 2003. *Smart Growth for Clean Water: Helping Communities Address the Water Quality Impacts of Sprawl.*

[51]U.S. EPA, Office of Water. March 2003. *Voluntary Guidelines for Management of Onsite and Cluster (Decentralized) Wastewater Treatment Systems.* EPA 832-B-03-001. March 2003. <www.epa.gov/owm/mtb/decent/download/guidelines.pdf>.

[52]U.S. EPA, Office of Wastewater Management. October 2000. *Potential Roles for Clean Water State Revolving Fund Programs in Smart Growth Initiatives.* EPA 832-R-00-010. <www.epa.gov/owm/cwfinance/cwsrf/smartgro.pdf>.

[53]Shiller, Gene. Southwest Florida Water Management District. June 27, 2003. Interview by Lynn Richards, U.S. EPA, Office of Policy, Economics, and Innovation.

[54]South Florida Community Development Coalition. Home page. <www.floridacdc.org/policy/tif2.htm>.

[55]State of Massachusetts Executive Office of Environmental Affairs. Community Preservation Act Web site. <commpres.env.state.ma.us/content/cpa.asp>.

[56]Trust for Public Land. Massachusetts Community Preservation Act. <www.tpl.org/tier2_rp2.cfm?folder_id=1045>.

[57]Rhodes, Milt. North Carolina Smart Growth Alliance. Email communication with Madelyn Carpenter, U.S. EPA, Office of Policy, Economics, and Innovation, June 27, 2003.

[58]Nebraska Environmental Trust. Home page. <www.environmentaltrust.org/index1.htm>.

[59]Clean Water Management Trust Fund. North Carolina Clean Water Management Trust Fund. <www.cwmtf.net>.

[60]Lacey, WA. Ordinance No. 1113. Municipal Research and Services Center of Washington. <www.mrsc.org/ords/L32o1113.aspx >.

[61]McElfish and Casey-Lefkowitz. 2001.

[62]Ibid.

[63]Charles River Watershed Association. Home page. <www.crwa.org>.

[64]State of California Legislative Counsel. Senate Bill No. 221, Chapter 642. <www.leginfo.ca.gov/pub/01-02/bill/sen/sb_0201-250/sb_221_bill_20011009_chaptered.pdf>.

[65]Sanchez, Rene. December 23, 2001. "New California Water Law Seeks to Curb Runaway Sprawl: Big Developments Must Show Ample Supply." *Washington Post.*

[66]U.S. EPA. Development, Community, and Environment Division. February 2003. *Using Smart Growth Policies to Help Meet Phase II Storm Water Requirements* [Draft].

[67]Washington State Department of Ecology. Stormwater. State of Washington. <www.ecy.wa.gov/programs/wq/stormwater/index.html>.

[68]Overeiner, Paul. September 2, 2003. "Officials develop initiative to improve Grand River." *Jackson Citizen Patriot.*

[69]U.S. EPA, Office of Wetlands, Oceans, and Watersheds. 2003. Fact Sheet, Water Quality Trading Policy. <www.epa.gov/owow/watershed/trading/2003factsheet.pdf>.

[70]Ibid.

[71]Environmental Trading Network Web site. <www.gltn.org/programs/cherry/cherry.htm>.

[72]Paulson, C.L. 1997. Testimony on the Cherry Creek Basin Water Quality Authority before the Water Quality Control Commission of the State of Colorado.

[73]Northeastern Ohio Areawide Coordinating Agency. Clean Water 2000: 208 Water Quality Management report Plan for Northeast Ohio. <www.noaca.org/Chapter-01__6017e_.pdf>.

[74]Openlands Project. October 2001. *Protecting Illinois' Environment through a Stronger Facility Planning Process.* <www.openlands.org/reports/FPA%20Report.pdf>.

[75]Campaign for Sensible Growth. October 2, 2002. Press Release. "Illinois EPA Preserves Facility Planning Areas." <www.growingsensibly.org/archive/archiveDetail.asp?objectID=1150>.

[76]Illinois Environmental Protection Agency. September 25, 2003. Press Release. "Advisory Group Selected to Oversee Pilot Testing for FPA Overhaul." <www.epa.state.il.us/news-releases/2003/2003-074-fpa-overhaul.html>.

[77]Northeastern Ohio Areawide Coordinating Agency. Clean Water 2000 Plan. <www.noaca.org/Chapter-01__6017e_.pdf>.

[78]Northeastern Ohio Areawide Coordinating Agency. Clean Water 2000 Plan. p. 16. <www.noaca.org/Chapter-11__6011e_.pdf>.

[79]Northeastern Ohio Areawide Coordinating Agency. Clean Water 2000 Plan. p. 11-4. <www.noaca.org/Chapter-11__6011e_.pdf>.

[80]Hernandez, Bobby. U.S. EPA Region 6. May 20, 2002. Interview by Lynn Richards, U.S. EPA, Office of Policy, Economics, and Innovation.

[81]Nonpoint Education for Municipal Officials. Home page. <nemo.uconn.edu>.

[82]The National Center for Smart Growth Research and Education. Education & Training. <www.smartgrowth.umd.edu/education/default.htm>.

[83]For contact information and a list of contractors certified in erosion control practices in Maine, see <www.maine.gov/dep/blwq/training/ccec.htm>.

[84]Maine Department of Inland Fisheries & Wildlife. Beginning with Habitat. <www.beginningwithhabitat.org>.

Photo courtesy of USDA NRCS.

SECTION II:

Site-Level Protection and Mitigation Measures

Where and how communities grow—directly and indirectly—affects water quality. As discussed in Section I, conventional postwar development patterns have had adverse effects on U.S. waterways. To help ensure the health of our watersheds, it is important to manage where growth occurs from a regional perspective. It is equally important to discuss how development should take place on targeted sites to reduce potential negative effects—the subject of this section.

In addition to regional water impacts caused by low-density, dispersed development, a number of site-level practices are detrimental to water resources. Setback and minimum lot size requirements maximize the amount of impervious surfaces around and between homes. Parking standards for shopping and office centers (as required either by localities or lenders) result in the vast parking lots that often characterize strip-shopping development. Zoning that separates uses (e.g.,

residential, commercial, office) often makes walking between destinations impractical, requiring use of vehicles that release emissions and toxic particulates that find their way to waterways through air deposition or polluted stormwater runoff. Some density restrictions forbid the construction of multi-story buildings or accessory units that could accommodate more units on less land.

Smart growth techniques provide a range of options for communities that seek a different approach to growth. Beyond the regional planning and coordination discussed in Section I, communities have also used smart growth approaches to improve site-level development. They have encouraged the development of existing impervious surfaces, in the form of infill development and brownfield and greyfield redevelopment. They have adopted a mixed-use, compact approach to site development that uses less land, and makes walking and other modes of environmentally friendly transportation feasible again. Communities have found that design considerations can not only improve the aesthetic quality of developments, but also their environmental quality. Finally, some communities are finding that smart growth techniques can actually provide greater flexibility for innovative developers. With this flexibility, developers are creating new construction and design that make sound economic and environmental sense, but are difficult or impossible to achieve under current laws.

EPA and other organizations, such as the Center for Watershed Protection, have written extensively about numerous BMPs and low-impact development techniques that reduce site- or development-specific stormwater runoff and associated pollutants.[85] When used in combination with regional techniques, these site-level techniques can prevent, treat, and store runoff and associated pollutants at the site. Many of these practices incorporate some elements of low-impact development techniques, such as rain gardens, biorention areas, and grass swales; many go further to incorporate smart growth principles, such as changing site design practices. Incorporating these techniques will not only help localities meet their water quality goals, but will also help create more interesting and livable communities. As with many development decisions, implementing these approaches could require communities to balance site-level impacts with regional benefits to achieve water quality improvements.

State and local governments can support improved site-level protection and mitigation measures through the policies discussed in the next four subsections: site planning, site-level technologies, ordinances and codes, and education. For the most part, policies described in this subsection support Smart Growth Principle #5: Foster distinctive, attractive communities with a strong sense of place. As in the previous section, issues to consider and practice tips are provided for many of the policies discussed.

SITE PLANNING

Local governments can direct development to specific areas within their communities. In addition, they can help plan for how that development occurs. This subsection focuses on planning approaches that help ensure development that is consistent with a community's smart growth and water quality goals.

For example, stormwater runoff varies substantially depending on a site's land use and design. Smart growth approaches can help communities prevent and manage their stormwater runoff and its effect on water quality and quantity. Overall site design considerations can have a dramatic impact on reducing stormwater runoff and associated pollutants.

In addition, critical ecological characteristics, such as steep slopes and permeable soil, also must be addressed when considering optimal site design to ensure that the design meets ecological and regional planning goals. Design and development practices that take into account the site's natural features can benefit water quality and support water quality improvements in the local watershed. Site design features, such as drainage and vegetation patterns, can increase onsite filtration of pollutants and minimize the impacts of site runoff on water quantity and quality.

Policy 47. Consider cumulative site-level development-related impacts

In most jurisdictions, only site-level impacts are considered in proposals for new developments. A more accurate assessment of development impacts, however, would consider the impacts from the current proposal as well as those of future planned and probable developments. Throughout time, the impacts from increased development across a region can have a compounding effect on regional water sources.

For example, limiting impervious cover at the site does not take into account the transportation-related infrastructure, such as roads and parking lots, or the retail venues that generally go along with development. Ten 100-acre sites that have 10 percent impervious cover will not simply translate into 1,000 acres with 10 percent impervious cover; the net increase in impervious cover will be much greater.

A better understanding of the cumulative water quality impacts of site-level regulation is necessary to ensure healthy regional water quality. Such an assessment would consider direct and indirect impacts, as well as short-term and long-term effects, resulting from current and proposed development. Having this cumulative information would allow local governments to better plan site-level development activities. For example, instead of limiting impervious cover at the site, they might wish to limit the total impervious cover within their jurisdiction.

Practice Tip: North Carolina's Department of Environment and Natural Resources developed a guidance document on cumulative and secondary impact assessment on aquatic and terrestrial wildlife resources and water quality. This document is intended to help local governments calculate the secondary and cumulative water impacts associated with public projects. The recommendations feature information on forested buffers, stream and wetland resources, infrastructure locations, floodplains, impervious surfaces and stormwater treatment, and erosion and sediment control. In addition, the guidance manual supports the development of model codes to further guide future construction.[86] These recommendations apply to new public developments and existing ones undergoing significant modifications or expansion.

Policy 48. Provide incentives to encourage specific development practices

A number of tools are available to communities to encourage development practices that serve smart growth and water quality goals. In addition to regulations mandating certain types of development, incentives can help shape development practices through voluntary changes. Incentives such as density bonuses, streamlined permitting, and decreased fees are all ways to reward development that incorporates features that improve water quality and enhance smart growth goals.

For example, a density bonus allows a developer to construct a building at a size and scale beyond that allowed by conventional zoning, thereby offering more opportunity for profit on the same amount of land. It is typically provided to developers as a reward or incentive when they provide a public amenity, such as parks, plazas, or affordable housing; stormwater benefits could also be included in the list of eligible public amenities. Municipalities also can offer decreased development fees for developments that include features to minimize impacts on waterbodies. Such features could include the use of pervious materials or landscaping that reduce runoff and treat water onsite. Bonuses or reduced fees can also be provided to developers who agree to replace older water and sewer infrastructure serving the project.

Local governments can create incentives to encourage landscaped setbacks and sidewalk medians. These features not only reduce runoff, but also improve the community's character.

This type of approach yields multiple stormwater benefits. More projects are likely to incorporate features that mitigate runoff, and the increased density allows more development to occur on less land, leading to more efficient use of existing roads, sidewalks, and water and sewer systems.

Photo courtesy of Local Government Commission.

Practice Tip: The city of Portland, Oregon, was the first in the nation to offer significant private sector incentives, in the form of density bonuses for developments that incorporate green roofs, to reduce runoff. In 2001, with a large concentration of new development along the Willamette River, the city approved the Floor Area Ratio bonus option for developments that include the use of landscaped rooftops to retain and filter rainwater. The program offers a sliding scale of density bonuses based on the size and relative scale of the green roof; developers can earn as much as three square feet of additional floor area for each square foot of green roof area.[87]

Photo courtesy of USDA NRCS.

Policy 49. Minimize stormwater runoff through construction site design

Construction activities are a major source of polluted runoff, especially sediments. Rainfall during the site development process leads to erosion from areas of bare soil left after vegetation is cleared and the site is leveled. Designing construction sites with sediment and erosion control in mind can minimize water quality impacts during construction.

Sediment in the street in Des Moines, Iowa, after a rain. Measures were not taken to protect the soil from erosion during development.

A key characteristic of smart growth communities is accommodating more residences, business, transportation, and retail uses on less land. During actual construction, using less land yields additional economic and environmental benefits for the simple reason that less land is required for the development; consequently, less soil is disturbed during construction, decreasing soil erosion and the costs for mitigating it. Further, the need for and expense of soil and erosion techniques, such as silt fences, are based on the number of acres disturbed. Building on fewer acres will save the developer money on soil and erosion technology. For example, a 1-acre site requires far less silt fencing than a 10-acre site, which calls for the same fence to be installed around its perimeter. If 10 residences are built on both sites, the per unit cost of erosion mitigation drops dramatically on the smaller site, demonstrating the cost savings that can be reaped through development of more compact sites.

Policy 50. Use conservation site design

Conventional site design typically divides available land into equal lots. In conservation design, lot division instead responds to the site's natural features, preserving large sections as open space and dividing the remaining land into smaller-sized lots for construction.

In its simplest form, conservation design (also known as cluster development) is development of a particular parcel in a manner that respects the site's natural and cultural features. Conservation design is usually applied to new residential developments in rural or suburban settings, where specific features—such as mature woodlands or existing trout streams—are preserved through a careful arrangement of new buildings and roads. These assets and other designated open spaces are often set aside for permanent conservation; building design and infrastructure concurrently take maximum advantage of these features (either as views or recreational sites).

WANT MORE INFORMATION?

The Minnesota Land Trust, with the University of Minnesota, developed a conservation design portfolio that highlights creative development options. It can be viewed at: <www.mnland.org/programs-consplanning.html>.

©2003 Regents of the University of Minnesota
All Rights Reserved. Used with the permission of Design Center for American Urban Landscape.

The homes at the Fields of St. Croix are clustered in blocks allowing 60 percent of the site to remain as permanent open space.

Available data demonstrate that conservation design in greenfield areas and in centrally located, compact, mixed-use developments has fewer environmental impacts because less land is required to accommodate the same number of units and commercial space than in low-density, dispersed developments. Conservation design benefits water quality by ensuring that large portions of new developments remain as permeable surfaces, with their ecological features intact. For example, open space preserved on the site can reduce runoff and allow infiltration of water to underground aquifers. Compact development techniques, such as clustering homes and buildings, reduce impervious surfaces.

Communities can encourage conservation design through open space zoning provisions that require developers to cluster density (e.g., residential units) on a site away from environmentally sensitive areas. Conservation easements could then be used to preserve the retained open space. Open space zoning is supplemental to conventional zoning and can be applied as an overlay district.[88]

Issues to Consider: Conservation subdivisions have become a popular tool to preserve open space. However, they should be used with care as they could lead to further separation of uses and increased dependence on automobiles. In some cases, conservation subdivisions can spur leapfrog development. In the context of a larger vision for the community, conservation subdivisions can play a vital role, but they should be avoided as a piecemeal tool or solution.

Practice Tip: The Jackson Meadow development in Minnesota incorporates typical conservation design principles. Located on a 145-acre parcel of high ground in open meadows and wooded hills overlooking the St. Croix River Valley, Jackson Meadow uses a cluster-housing model, preserving more than 70 percent of the site as open space. Housing and street patterns reflect existing models in the nearby town of Marine, and the development is organized topographically with neighborhoods oriented toward a central green. In lieu of typical suburban streets, each neighborhood block shares a pedestrian way located between the fronts of houses. The site is connected to Marine through a series of walkways and pedestrian corridors linked to the central green. Each pedestrian way connects directly to more than 5 miles of walking and cross-country skiing trails. From these trails, residents of Jackson Meadow are within a 10-minute walk of the local elementary school and Marine's downtown village center. This new neighborhood highlights the importance of walking, sustainability, and diversity, and designating the best land as open space for community interaction and recreation.[89]

Policy 51. Minimize stormwater runoff through traditional and non-traditional BMPs

While BMPs are accepted practices to reduce stormwater runoff, numerous opportunities exist within the BMP framework to employ "non-traditional" smart growth practices to reduce stormwater runoff and associated pollutants.

Communities can expand the concept of BMPs by incorporating "non-traditional" approaches into their environmental management practice to reduce stormwater runoff to its lowest possible levels. These approaches might include using compact site design, preserving open space, incorporating street trees into a site design, requiring planters within plazas, or improving comprehensive planning. Such strategies not only reduce runoff but also foster distinctive, attractive communities. This type of multi-objective approach is central to smart growth.

> **Practice Tip:** The state of Maryland has developed the Maryland Stormwater Design Manual, which includes both design standards and environmental incentives. The manual aims for better stormwater management by relying less on standard BMPs for all development projects and more on an approach that mimics existing hydrology through site design policies. The goal is to protect the state's waters from adverse impacts of stormwater runoff, provide design guidance on the most effective structural and non-structural BMPs for development sites, and generally improve stormwater management practices on development sites in the state.[90]

Policy 52. Designate smart growth site design as a BMP

EPA provides a menu of onsite BMPs to reduce stormwater runoff.[91] As discussed in the previous policy, a number of non-traditional BMPs help reduce runoff, decrease associated pollutants, and enhance the look and feel of a neighborhood. Designating a smart growth site design deserves special mention in an expanded policy toolbox because of its potential to minimize development-related water quality impacts.

WANT MORE INFORMATION?

The Center for Watershed Protection maintains a Web site with information and resources for people involved in stormwater management. The site is located at: <www.stormwatercenter.net>.

To designate a site design, regulators should identify and define criteria for numerous design principles, including density levels, the number of uses the site accommodates, percentage of open space—including plazas, social gathering areas, or other public amenities—and the range of transportation and housing choices available. Individually and collectively, these design features reduce overall land consumption and impervious surface compared to more conventional development designs.[92] Designating smart growth site design as a BMP is an option at the state or municipal level, providing another tool for developers to use to reduce stormwater runoff and associated pollutants.

At the state level, smart growth site design could be designated as a BMP where land use controls are explicitly stated, such as within the state's general permit, any stormwater management guidelines, or model stormwater ordinances. In addition, although general permits in most states do not include specific suggestions on how localities can manage their stormwater runoff, they do include sections that require minimum control measures. States could include a section on reviewing or considering site designs within the permit approval process, recognizing the importance of site design in managing stormwater runoff.

At the municipality level, several opportunities are available for specifying smart growth site design as a BMP. A municipality can adopt a stormwater ordinance that includes smart growth or modify existing ordinances to ensure that they allow developers to use a smart growth site design as a BMP or to receive some other type of water quality credit. In addition, municipalities can designate a smart growth site design BMP as part of their public facilities manual, which provides a blueprint for developers on how to implement ordinances and other local requirements. By defining and establishing specifications for a smart growth site design within this manual, the municipality supports developers with the information they need to design and build smart growth communities.

Policy 53. Allow green building points for infrastructure repair

Green buildings are growing more popular as localities realize the benefits of buildings that use less energy, contain better materials, and treat stormwater on the site. In older cities and suburbs, however, site constraints such as the existence of legacy pollutants, sewer and water pipes that are failing or in disrepair, and expensive land often limit or prevent a developer's ability to follow standard green building practices for infiltrating stormwater on the site. A certified green building program could award points for infrastructure repair. These infrastructure repairs can encourage additional development activity in areas needing revitalization.

For older cities, water and sewer pipes in disrepair can be a significant water quality issue. During heavy rains, overtaxed sewer lines back up into homes and streets with stormwater and sewerage. Leaky water pipes mean that cities pay for water that seeps into the ground rather than being delivered to customers. A city with a green building scorecard could add a category for developers who want to replace or repair the failing water and sewer infrastructure serving, or proximate to, their projects. These "innovation points" would have to be tied to the project and be awarded based on repair of an identifiable source of water problems.

> **WANT MORE INFORMATION?**
>
> The Green Building Council sponsors the Leadership in Energy and Environment Design (LEED) scorecard, which is a popular tool for localities that want to reward developers who follow green building designs. Information on LEED standards can be found at: <www.usgbc.org/LEED/LEED_main.asp>.

Policy 54. Allow offsite mitigation

Current approaches to stormwater management generally require onsite practices, such as detention ponds. These approaches might not always be practical, however, in higher-density areas or in compact, mixed-use communities. Another approach to ensuring that stormwater is effectively managed is to allow offsite mitigation.

Offsite mitigation allows a developer to treat stormwater runoff at another location, specified by the local government, in lieu of treating runoff at the development site. Localities must approve the project in advance and ensure that it conforms to existing building and zoning regulations and provides for long-term site maintenance. Offsite mitigation provides an opportunity to strategically locate mitigation facilities where they can better address existing or potential water quality issues.[93] For example, Nashville, Tennessee's stormwater ordinance states, "if it is unfeasible to implement onsite stormwater BMPs, then the development could design a system that controls quality for an equivalent portion of runoff entering from the watershed above."[94]

In return for offsite mitigation, jurisdictions may increase allowable densities in downtown and designated areas, for example, and then assume responsibility for maintaining water quality in that particular area. This strategy allows developers to build communities that integrate residential, commercial, and transportation uses—and the resultant runoff flow—into the community and offset their water impacts elsewhere, thereby ensuring overall regional water quality.

WANT MORE INFORMATION?

More information about the CWP's Roundtable series, smart site practices, and better site design techniques, is available at: <www.cwp.org/smartsites.pdf>.

Policy 55. Adopt model development principles

Sometimes development strategies that preserve open space and minimize impervious cover are practiced in some municipalities but not others nearby, undermining efforts to improve overall regional water quality. Communities or organizations can support more widespread adoption of improved development by adopting model development strategies that minimize impact on water resources.

Existing planning and zoning regulations prescribe many of the features of conventional development, such as large surface parking lots and dispersed, low-density developments that adversely affect water quality. Using alternative development design often requires time to obtain a zoning or other regulatory exemption—a time-consuming and costly process. As a practical matter, widespread implementation of development strategies that preserve open space and minimize impervious cover requires fundamental changes in the framework that determines how and where land is developed. Such fundamental change requires a comprehensive community approach that identifies key priorities and coalesces in a shared vision of the type of future growth that is desirable. Clear policy guidance, in the form of model development principles, could be drafted and adopted by local jurisdictions to help the community achieve its goals.

In 1996, the Center for Watershed Protection (CWP) began a project that provides an example of how to carry out this process. Recognizing the link between site design and watershed health, the CWP initiated a "Site Planning Roundtable" to encourage better design at the site level. In the first phase of this national-level project, a roundtable group consisting of planners, engineers, developers, attorneys, fire officials, environmentalists, and transportation and public-works officials from nationally recognized organizations came together to develop and endorse a set of national model land development principles. Meant to promote economically viable and environmentally sensitive site planning, these principles include the following[95]:

- Shorter, narrower streets
- Smaller parking lots
- Increased stormwater treatment practices
- More community open space
- Increased vegetated buffers
- Enhanced native vegetation
- Limited clearing and grading

Arlington County, Virginia, decided in the mid-1980s to encourage high-density development around transit stops in order to maintain the neighborhood feel of surrounding lower-density communities.

Photo courtesy of U.S. EPA.

Starting with these principles, numerous communities have since conducted their own site planning roundtables, in which local stakeholders review the CWP template and adapt it to include the principles that make sense for their own communities. These roundtables aim to provide communities with a technical and economic framework to rethink their zoning and subdivision ordinances, planning processes, and individual site development decisions. By strategically helping communities revise their planning and zoning ordinances and incorporate model development principles, such projects provide local governments with the tools to promote more environmentally sensitive development across the entire region.

Practice Tip: The Frederick, Maryland, roundtable project adapted design principles developed at the national level for local application. The Frederick County Site Planning Roundtable was initiated partly as a result of conversations between the county's planning and zoning staff and CWP staff. Employees of CWP had observed that the county was rapidly developing using conventional practices because many of the county's codes actually prohibited more innovative development strategies that would reduce impervious cover. Using a consensus-building process, the project identified local codes and ordinances that prohibited or impeded better site designs. Roundtable members representing a wide range of professional backgrounds were invited to participate in a nine-month process to review the county's existing subdivision and zoning codes. The roundtable reviewed the model development principles to identify which modifications were needed for application to Frederick County and summarized its findings in *Recommended Model Development Principles for Frederick County, Maryland.*

WANT MORE INFORMATION?

Frederick County summarized its findings in *Recommended Model Development Principles for Frederick County, Maryland,* available at: <www.cwp.org/Frederick.pdf >.

Policy 56. Allow developers to pool stormwater management efforts

Traditionally builders or developers are responsible for stormwater management efforts only on their particular sites. Smart growth suggests another approach—allowing developers to work together and pool resources and strategies for joint stormwater management efforts. Such joint efforts can yield better environmental results and can also achieve cost savings. Moreover, allowing developers to pool stormwater management efforts can provide more flexibility for the developers working in space-limited areas, such as infill sites. To encourage urban revitalization efforts, infill development, and other development scenarios that might be space-limited, communities could implement more flexible regulations for site-level mitigation that would permit developers to work together and pool resources for handling stormwater.

Practice Tip: San Diego, California, has introduced flexible regulations to allow the developers of multiple properties within infill development areas to pool their resources for handling stormwater. Rather than requiring each property to implement BMPs, the new rules allow developers to contribute to larger basin-wide controls that serve a cluster of redeveloped properties. This method is called the "localized equivalent area drainage" method. The city believes treatment systems with a larger capacity serving a cluster of properties can remove the same amount of pollutants as individual devices, such as filters placed where water enters storm drains. By pooling resources, the city estimates that developers will save up to $40,000 per acre.[96]

SITE-LEVEL STRATEGIES AND TECHNOLOGIES

The previous subsection focused on site planning approaches that communities can implement to ensure development consistent with their smart growth and water quality goals. This subsection describes strategies and techniques for the site design process of a particular development. These strategies can help communities achieve their goals based on how they want their neighborhoods to look, act, and connect with other neighborhoods and still meet water quality objectives.

Policy 57. Maximize use of existing impervious cover

Redevelopment of previously developed sites provides water quality benefits by reducing the need to accommodate growth on undisturbed, open land. These benefits increase when the redevelopment of a site maximizes the use of already impervious cover by modifying it to serve multiple uses.

It is well known that the amount of impervious cover in a watershed directly affects the volume of runoff, contributing to higher pollutant loads, more frequent flooding, and the degradation of stream channels. As discussed previously, redevelopment of brownfield or greyfield properties can decrease runoff. The logic behind this phenomenon is simple: a parking lot that was previously 100 percent impervious cover will have close to 100 percent runoff. Changing the use of that land by adding houses, apartments, retail, or pocket parks will not increase runoff, but will, in most cases, decrease it. In addition to brownfield and greyfield opportunities, many communities might have smaller sites of existing impervious cover that could accommodate redevelopment activity. These more common opportunities include vacant and abandoned buildings, land that held property that has since been torn down, underutilized retail areas such as declining strip malls, or out-of-business gas stations. Identifying and marketing these properties as potential places for redevelopment will not only help revitalize neighborhoods, but will reduce the need to accommodate growth on undisturbed land.

In addition, many impervious surface areas can be redesigned to capture runoff or otherwise made to serve more than one use. By assessing and taking advantage of such possibilities, communities can reduce runoff from impervious surfaces, such as parking lots and rooftops. For example, rooftops that previously contributed to runoff volume could be redesigned to capture and direct water to landscaping uses. Plazas that serve as gathering places for lunchtime workers might, for example, serve double duty as overflow parking lots for evening or weekend area visitors. Underground parking, shared parking, and multi-purpose parking lots (including those that serve as sites for markets or recreational facilities in off-hours) all serve to eliminate the redundancy of facilities and reduce the need for construction of additional impervious surfaces.

Policy 58. Design open space areas to minimize stormwater runoff

Incorporating small areas of open space, such as plazas or pocket parks, within compact developments can serve a number of critical functions: as a gathering place for residents, as a focal point for the development, as a tool to encourage privacy and division of spaces, and as an environmental resource. With some strategic design modifications, these valuable open space resources can often be used to reduce stormwater runoff and still serve to create more attractive, distinctive communities.

Lawns can be modified to capture and treat runoff.

Many redevelopment and infill projects use open spaces, courtyards, and plazas to provide a community focal point, encourage community interaction, and offer opportunities for recreation. Often they consist of large areas of impervious surface, such as great swaths of concrete or large circulating fountains. Others are comprised of landscaping features that support infiltration and water retention. Communities can reduce overall imperviousness by encouraging developers to expand their use of landscaping and alternative covers—such as pavers, bioretention areas, or planting boxes—that allow for water infiltration. These materials can often support the same functions as their impervious counterparts and also serve to store, filter, or treat rainfall to reduce the impact of runoff on water resources.

Photo courtesy of USDA NRCS.

Practice Tip: The Buckman Heights residential development in Portland, Oregon, captures and filters rooftop runoff through a centralized courtyard featuring two gardens of native and ornamental plants. A third vegetated channel is located adjacent to the parking lot. The soil and plants in these gardens act as a natural filter and reduce stormwater runoff. In addition, narrower driveways and the use of a back-up dry well reduce the amount of runoff generated. These combined efforts allowed the site to be built without connection to the stormwater system and ensured that the development will not contribute to the city's stormwater treatment needs.[97]

Policy 59. Preserve and enhance green areas in existing neighborhoods

In many cases, vegetated areas remain in existing neighborhoods, community parks, abandoned properties, or natural areas such as non-recreational streams or lakes. Such areas make positive contributions to a community's water quality through infiltration or reduced imperviousness, but they are often fragile assets, small and fragmented, and strongly influenced by adjacent uses. Often they are susceptible to compaction, dumping, and invasive plant species from adjacent developed sites.

Careful management of fragile or damaged green areas will encourage revegetation and soil restoration and contribute to more attractive communities with a strong sense of place. In approaching these publicly owned or abandoned sites, communities are advised to consider the type of vegetation most likely to improve water quality. For example, grass-covered sites are less likely to filter water and mitigate runoff from neighboring sites than those with native vegetation. Lawn grass is generally compacted during its installation and remains so during maintenance (e.g., continual mowing). Communities must balance the need for water quality improvements with the specific requirements called for by the site and its surrounding residents and uses. In addition, thoughtful planning and zoning for developed uses in the vicinity of these sites can also help to mitigate impacts upon these resources and ensure that they provide important community and water quality benefits far into the future.

Conservation easements, donations of public land-to-land trusts, and innovative partnerships for the care of land (such as between a nearby association or school and the local jurisdiction) are among possible long-term solutions for financing and maintaining these sites. By whatever mechanism they are managed, attractive and well-maintained green spaces can serve as community assets, spurring more investment and redevelopment of the surrounding areas.

Practice Tip: The 26th Street Gateway in Philadelphia, Pennsylvania, was previously a post-industrial wasteland of neglected spaces, crumbling asphalt, and short-dumping sites. In 1989, the organization Philadelphia Green joined with public and private organizations (including the Pennsylvania Department of Transportation and Philadelphia's Department of Streets) to rehabilitate the stretch of roadway. Natural areas were preserved, and native vegetation was planted. Now this 1-mile stretch of land covering 25 acres is a meadow of native trees, grasses, and wildflowers.[98]

Policy 60. Use green practices to manage rooftop runoff

Rooftops are by necessity built with impervious materials such as asphalt, metal, shingles, and other tiled materials. They can still provide an effective means of reducing runoff from sites, however, particularly in higher-density areas, if practices such as rooftop gardens and other green infrastructure practices are used.

Rooftop runoff can be managed through the storage, reuse, and redirection of runoff for stormwater management and other environmental benefits. Green roofs, in which some or all rainwater is absorbed and redirected to other uses (such as rooftop gardens), can be used to reduce the volume of rooftop runoff. Gutter systems can be designed to direct runoff from roofs into rain barrels, which subsequently provide a "grey water" resource for landscaping and thereby reduce water demand. Runoff volume can also be reduced through improvements in the design of rooftops and site layout, so that the reduced flow from less sloped roofs is directed onto pervious surfaces instead of into stormwater systems.

Such techniques are useful in lower-density development, yet they also have particular significance in higher-density, compact developments where marginal per unit decreases in runoff become significant when multiplied by the greater number of units located onsite. These cumulative effects might be great enough that they eliminate the need for detention ponds or other mitigation efforts that might otherwise interrupt the flow and feel of a compact community. In addition, such mitigation efforts can help communities avoid hotspot effects. Further, any effort to reduce the pressure on an overtaxed stormwater infrastructure means that more growth must be accommodated in existing neighborhoods, so that open land on the urban fringe can be preserved.

WANT MORE INFORMATION?

Increasingly, cities, private industry, and residents are installing environmentally friendly roofs. A wide variety of case studies, information, and technical resources are available at: <www.greenroofs.com> and <www.cleanrivers-pdx.org/clean_rivers/ecoroof.htm>.

Rooftop runoff can be directed to backyard ponds.

Photo courtesy of USDA NRCS.

Issues to Consider: Specially trained architects must be employed to design systems that do not overwhelm the structural capacity of the roof, and to ensure that the appropriate types of vegetation are used in a manner that is both cost-effective and protects the rooftop's sustainability and its stormwater management capabilities.[99]

Practice Tip: Completed in the spring of 2001, Chicago's City Hall rooftop garden covers approximately 20,300 square feet and contains a variety of grass, shrub, vine, tree, and other plant species. The roof's water storage slows down and reduces direct discharge into storm sewers, resulting in less pressure on the sewer system and improved water quality. The green roof is cost-effective, generating direct energy savings through a combination of shading, evapotranspiration effects, and insulation.[100]

Policy 61. Use low impact development techniques

Low-impact development (LID) techniques are those that mimic the predevelopment site hydrology to store, infiltrate, evaporate, and detain runoff. They are a natural complement to smart growth approaches that seek to reduce runoff through an improved approach to regional development and site design. Although smart growth approaches applied at the site level reduce the volume of runoff, the use of LID techniques adds to the potential gains by mitigating the effects and pollution levels of the site's stormwater runoff.

LID techniques are usually associated with new development sites, such as subdivisions, parking lots, or other large uses with a high level of imperviousness, and where the hydrological and topographical aspects of the site can easily be determined. Some aspects of the LID approach, however, are equally applicable to and potentially beneficial for infill development. For example, vegetated buffers can be located next to sensitive areas such as streams to slow the movement of runoff and filter sediment and pollutants. Level spreaders are site features that convert concentrated runoff (such as that from a pipe that carries runoff from a number of impervious surfaces) to sheet flow that can be more evenly dispersed across a slope, thereby causing less erosion than a single, high-volume stream.[101]

Jordan Cove, a low impact development in Waterford, Connecticut, uses rain gardens between houses.

The potential for using LID techniques for urban infill areas is increasing. Ongoing research is being conducted to evaluate the impact of LID techniques in urban settings, as compared to their

Photo courtesy of the NEMO program and the University of Connecticut.

traditional application in rural and suburban contexts. More research is needed to better understand the quality and quantity of runoff under various redevelopment approaches and the potential economic savings to be gained by using LID to capture stormwater flow before it enters a system that is at or over capacity.

Issues to Consider: Communities must resolve the question of how to pay for LID features on a site. Given that reduced and/or improved stormwater runoff can mitigate the need for treatment cost and system expansion, it might be appropriate to offset the costs borne by private developers who incorporate LID through some financial incentive, such as reduced fees. It might also be determined that the aspects of LID that serve to reduce conventional site development costs— such as clearing and grading—might be sufficient to offset any higher costs for constructing features such as those discussed above. Further, the long-term cost savings (in terms of turf and pavement maintenance and replacement) that are generated by LID features could convince private developers that the additional investment in stormwater mitigation site technology is worthwhile.

WANT MORE INFORMATION?

The Low Impact Development Center offers a range of technical information, resources, and tools at: <www.low impactdevelopment.org>.

Practice Tip: In the Puget Sound area of Washington State, King County officials have merged their LID program with the community's larger smart growth initiative to develop comprehensive planning and implementation for stormwater management. The Puget Sound Action Team, comprised of community leaders, local governments, tribes, and businesses, oversees water quality protection in the sound by setting up work plans and implementation goals for involved groups. Projects to date include a LID CD-ROM with materials from the LID in Puget Sound Conference, and an Alternative Futures project with the public to test alternate land use scenarios with hydrologic and habitat models.[102]

Policy 62. Construct narrow, walkable, well-connected streets

Many development sites today are connected by wide streets made of large quantities of impervious surface. The increased street width is not needed in all instances and can make unpleasant, inconvenient, and at times unsafe places to walk. Impervious surface can be reduced and walking can be encouraged if site design incorporates narrower, walkable, well-connected streets for both vehicles and pedestrians to use. As a result, runoff can be reduced and air and water quality improved through the reduced need for vehicular transportation.

Photo courtesy of U.S. EPA.

Downtown Annapolis, Maryland, demonstrates that narrow streets can still provide on-street parking, which serves as a buffer for pedestrians.

Communities can express their preference for reduced runoff from narrower streets that are better connected and use less impervious surfaces through design guidelines. Site design guidelines might also call for alleys or rear lanes that serve multiple functions, such as utility and service areas, thus better maximizing the use of existing impervious surfaces. Some counties and metropolitan planning organizations have clarified their objectives for street design in formal street design guidelines. Others have stated a maximum level of impervious surface for a particular parcel or watershed, and then give developers and designers flexibility to meet runoff reduction requirements using a variety of techniques, including open space, narrow roads, parking structure design, and reduced building footprint. North Carolina's Department of Transportation, for example, approved street design guidelines to make it easier for local governments to implement traditional neighborhood street networks in new developments. The guidelines specify street width and require bicycle and pedestrian facilities, which support improved water quality as well.[103]

Issues to Consider: One critical component of a community's transportation system is effective emergency response; fire, ambulance, and police officials need to respond to calls quickly. To meet this need, roads are built to accommodate large fire trucks with large intersections for faster turns. In some instances, communities have abandoned plans for smart growth road and transportation improvements, such as multi-use streets or engineering techniques to calm traffic, after fire chiefs testify against the plans based on faster response times. Some emergency response officials have pointed out, however, that the wider streets and turns actually produce more safety problems than they solve, since they allow for higher speeds for all traffic. Others note that residential street designs, such as cul-de-sacs and limited access points for private communities, also impede effective response times. To achieve safer street networks, local governments should consult emergency responders during the design phase of a road improvement project, rather than at the end of the process. They should identify street and traffic solutions that work well for everyone.

Practice Tip: The city of Columbus, Ohio, has developed a stormwater ordinance that supports the reduction of impervious surface—including narrower street widths that conform to the standards found in the Traditional Neighborhood Development code—to lessen the impacts from runoff. Other strategies include a reduction in commercial parking and the preservation of open space, including agricultural lands and riparian areas.[104]

ORDINANCES AND CODES

Ordinances and codes are means by which a community can express its goals and objectives for development. Ordinances and codes help shape the type and placement of development in a community and manage its natural resources. As such, they can be used to promote standards to better manage how and where development takes place.

Policy 63. Adopt stormwater ordinances

WANT MORE INFORMATION?

The Center for Watershed Protection maintains a Web site containing model stormwater ordinances at: <www. stormwatercenter.net>.

Local governments are currently not required to have stormwater ordinances in place. Adopting such an ordinance, however, is advisable because it lets communities effectively enforce development and mitigation guidelines that protect water quality by reducing the quantity or improving the quality of stormwater runoff.

Stormwater ordinances give local governments the legal authority to shape development and better protect water quality. The adoption of enforceable stormwater ordinances is critical to implementing new and innovative ways to prevent or control stormwater runoff. Such ordinances can require developments to conduct regular maintenance activities. For example, local governments can set surface runoff limits for post-construction stormwater runoff volumes and identify allowable nonstructural and structural stormwater practices. The ordinances can also include language regarding onsite stormwater requirements, and whether offsite treatment is an option.

State and regional governments can support communities by developing model ordinances that can be customized to a locality's conditions and preferences. The model ordinance developed by the Twin Cities Metropolitan Council in Minnesota, for example, includes design standards for stormwater ponds, BMPs for protecting water quality, and shoreline regulations.[105]

WANT MORE INFORMATION?

EPA offers a range of tools and examples of stormwater ordinances on its Web site at: <www.epa.gov/owow/nps/ordinance/stormwater.htm>.

Issues to Consider: Stormwater ordinances are most effective when they clearly identify the entity responsible for long-term maintenance and build in a requirement for regular inspection visits. Ordinances might call for the use of BMPs; they should also provide supporting information, such as maintenance agreements and inspection checklists, to ensure that they result in the desired water quality impacts and perform efficiently during the long term. In addition, ordinances must be comprehensive enough to ensure that regional water benefits are achieved, but specific enough to reflect the needs of particular areas. Older urbanized areas, for example, will face different stormwater issues than new developments.

Practice Tip: Grand Traverse County, Michigan's Stormwater and Sediment and Erosion Control Ordinance is an example of an ordinance specifying operation and maintenance provisions for stormwater, erosion, and sediment control. The ordinance specifies actions property owners must take, including certification that construction has been completed in accordance with the approved soil erosion and stormwater runoff control plan, inspection procedures, and other compliance and enforcement actions regarding stormwater, sediment, and erosion control.[106]

Policy 64. Adopt ordinances for source water protection

Under the Safe Drinking Water Act (SDWA), all states are required to complete assessments of their public water systems that delineate areas that feed groundwater and surface water supplies, and identify potential pollution risks. Additionally, to further ensure water quality, a limited number of communities have ordinances in place to protect source water. Communities should consider developing ordinances that protect source waters, such as aquifers and watersheds, by adopting ordinances that protect the most critical recharge or contribution areas, nearest to wells and intakes.

The purpose of source water protection is to prevent pollution from reaching the groundwater, lakes, rivers, and streams that serve as local communities' drinking-water sources. Ordinances can be developed to protect water sources and help safeguard community health by reducing the risk of contamination of water supplies. Wellhead protection zones and aquifer protection areas are two examples of source water protection ordinances that help protect groundwater sources. Water supply watershed districts and lake watershed overlay districts are examples of local management tools that provide protection to surface water supplies by restricting land uses around a reservoir used for drinking water. In all cases, communities can develop an ordinance that applies to a specified area surrounding the water source in question. Such ordinances are most effective when they provide clear guidance on the allowable uses, water quality measures required during construction or in existing developments, and other practices to help protect and ensure the quality of the community's drinking-water sources.

Issues to Consider: Source water planning should be conducted on a scale that ensures protection of the entire recharge zone for that particular water source. It is unlikely that communities will be able to protect, or perhaps even define, entire recharge zones, as these zones can be very large and could include substantial areas outside of a community's jurisdictional boundaries. For surface waters, communities might wish to create overlay zoning districts that have boundaries large enough to protect the source water resource, tributaries, and the contributing streams.

For groundwater protection, communities can consult with the U.S. Geological Survey (USGS) to ensure that their overlay zoning district encompasses the entire area that recharges an aquifer. In addition, communities could contact the state agency responsible for source water assessment. Many states have completed a comprehensive effort to delineate and characterize critical wellhead protection and surface water contribution areas for every public water system.

In addition, an ordinance should include specific information on the allowable and prohibited land uses within the source water protection zone. For example, many source water protection ordinances limit or forbid the storage of hazardous materials and place restrictions on the location of businesses that use these materials within the district. An ordinance should include procedures for the review of proposed projects within a source water protection district to verify that the project is consistent with the ultimate goal of the ordinance. These procedures might include requiring applicants to submit geotechnical and hydrological analyses to determine the potential impacts to water quality, and the submission of spill control plans for businesses performing potentially contaminating activities. Finally, the ordinance should include language explaining the mechanisms for enforcement of the ordinance, including the civil and criminal penalties that could apply for failure to obey. Local governments might wish to review state statutes and regulations governing municipal land use and talk with public health authorities, to assure consistency and avert concerns regarding state preemption.

WANT MORE INFORMATION?

A new EPA source water protection rule, Long Term 2 Enhanced Surface Water Treatment, allows treatment credit for watershed protection actions. Details are available at: <www.epa.gov/ogwdw/lt2/index.html>.

WANT MORE INFORMATION?

EPA's Office of Water has numerous resources on planning and implementing source water protection programs, including financial assistance, case studies, and model ordinances available, at: <www.epa.gov/safewater/protect/sources.html>.

Practice Tip: The New York City Watershed Agreement provides a dramatic example of communities taking steps to protect their source water. In 1997, EPA and New York City, along with more than 70 towns and eight counties, signed an agreement to support an enhanced watershed protection program for the New York City drinking-water supply. Through the multi-year, $1.4 billion agreement funded by the city, a multi-faceted approach is being implemented, including the purchase of 80,000 acres within the watershed to protect drinking-water sources. This plan allows the city to avoid the construction of filtration facilities estimated to cost between $6 billion and $8 billion.

This agreement created a blueprint for protecting the watershed during the next 10 to 15 years and established a land use pattern intended to protect the future of the city's water supply. The city has clearly demonstrated a commitment to the protection of the watershed through the provision of green infrastructure in established villages, economic development aid to bolster a healthy rural economy and working landscape, and support for various planning studies.[107] These efforts serve to correct existing water quality problems, prevent development in important ecological areas, promote pollution prevention, and create and strengthen organizations and local governments in their ability to manage growth and protect water quality.

Policy 65. Adopt water-saving landscaping ordinances

In addition to its many environmental benefits, smart growth fosters the development of distinctive, attractive communities with a strong sense of place. Landscaping ordinances adopted at the local level can serve this function and provide water quality benefits when they encourage the use of water-saving landscaping or xeriscaping™.

Communities can foster distinctive places and achieve water quality benefits by adopting ordinances that call for the use of native species, especially perennials, in landscaping. Such plants can reduce water use because they are well adapted to the climate and therefore require less water and maintenance. An ordinance might encourage the expanded use of xeriscaping—an approach to landscaping that relies on the use of plants and landscaping techniques that explicitly reduce water use. This type of landscaping approach tends to provide more permeable surfaces than conventional landscaping, thus further reducing stormwater runoff.

Issues to Consider: Some planned communities use neighborhood covenants to regulate the type of landscaping in their community to ensure consistency in appearance. In extreme cases, they might ban xeriscaping and prescribe the use of a specific, water-thirsty type of groundcover, such as Kentucky bluegrass. One community is seeking to remove these bans by opposing a proposed law that would forbid new subdivisions in Denver, Colorado, from requiring landscaping and banning the use of xeriscapes. Denver officials want more homeowners to consider landscaping techniques that feature plants that require less water, but sometimes are viewed as unappealing by neighbors.[108]

Practice Tip: Florida's water management district rules require that all local governments consider adopting a xeriscape ordinance as a water conservation measure. The Florida DEP prepared a model landscape ordinance that minimizes irrigation and uses landscaping to protect water quality. The ordinance would apply to all new construction and sites undergoing renovation that require a local building permit.[109]

Policy 66. Adopt tree ordinances

Tree ordinances are among the many ways localities can foster distinctive, attractive communities that also achieve water quality benefits. By encouraging communities to plant more trees, tree ordinances help achieve these dual goals.

The stormwater benefits that trees provide are often not fully recognized. Trees intercept and slow the fall of rainwater, helping the soil to absorb more water for gradual release into water sources. This cycle prevents flooding, filters out toxins and impurities from the water, releases water into the atmosphere, and reduces stress on the stormwater system. Based on these various benefits, developers and residents should be encouraged to plant and maintain trees.

Tree ordinances are most effective when they specify the goals of a community's tree program, its methods of enforcement, and evaluation procedures. In addition, they should provide clear guidelines and rules on how to plant and manage new and existing trees on new development sites and along public streets. For example, street tree ordinances can explain the practice of planting and removing trees within the public right-of-way. They might also specify planting requirements for parking lots, thereby mitigating the effects of their imperviousness. Smart growth projects and developments can be designed to maximize the preservation and use of trees to help improve the quality of a community's water resources.

Volunteer programs, such as AmeriCorps, can assist in implementing a community's tree ordinance.

Photo courtesy of USDA NRCS.

WANT MORE INFORMATION?

American Forests developed a software package called "CITYgreen," which can help establish a baseline tree canopy and estimate the dollar value of the services provided to a community by its tree cover. Garland, Texas, used CITYgreen to measure the cost savings associated with its tree canopy and learned that its trees provide 19 million cubic feet in avoided stormwater storage space, saving the city an estimated $2.8 million annually in construction costs for a stormwater facility. This tool is available at: <www.americanforests.org/graytogreen/stormwater>.

Issues to Consider: Different trees have different absorption rates, growing condition needs, growth rates, and lifespans. Policymakers should consult an expert to determine which trees will provide the most water quality benefits for the community. In addition, planners should ensure that the trees' future needs are met by ensuring that tree planters are large enough to support tree growth in the coming years.

Policy 67. Implement ordinances and standards to better manage development along waterways

Waterbodies are particularly sensitive to the uses that surround them. Polluted runoff, construction sediment, and the elimination of natural features that filter water can have a dramatic effect on the quantity and quality of water resources. Communities can develop and implement riparian standards and buffer ordinances to protect zones along and around waterbodies. Furthermore, by preserving and maintaining the land surrounding waterbodies, the community's character can be enhanced.

Riparian standards can help minimize the impact development has on riparian zone functions by better directing and managing development. To be effective, standards should consider the particular characteristics of the riparian zone and waterbody being protected. For example, a small spring-fed creek will have different requirements for protection and accommodate different nearby uses than will a man-made lake. Riparian areas have high ecological value, and standards designed to protect them are critical to ensure that future development does not pose further threats.

Buffer ordinances, which protect water quality and aquatic habitat, regulate activity in the strips of native vegetation along streams and other water resources. These areas provide wildlife habitat, protect water quality, and serve as natural boundaries between local waterways and existing development. Buffers help protect water resources from the impacts of development by filtering pollutants, sediment, and nutrients from runoff. Other benefits of buffers include flood control, stream bank stabilization, stream temperature control, and room for lateral movement of the stream channel. Ordinances can specify the size and management of the stream buffer.

Issues to Consider: To provide the functions necessary to protect water resources from the impacts of development, buffer ordinances should require that buffer boundaries be clearly marked on local planning maps. In addition, language should restrict vegetation and soil disturbance during maintenance, tables should illustrate buffer width adjustment by percent slope and type of stream, and direction should be provided on allowable uses and public education.

Practice Tip: The state of Maine created a Mandatory Shoreland Zoning Law that requires municipalities to protect shoreland areas by zoning land within 250 feet of coastal waters, lakes, and rivers, and within 75 feet of second-order perennial streams. These zoning ordinances provide guidance on the types of activities that can occur by establishing zones for resource protection, general development, residential, and other uses, and by specifying building size and setbacks for those areas in which development will occur. In addition, Maine's revised Natural Resources Protection Act (NRPA) regulates development activity within 75 feet of any mapped stream. To receive an NRPA permit, applicants must demonstrate that the proposed activity will not cause unreasonable erosion of soil or sediment or prevent naturally occurring erosion; unreasonably interfere with the natural flow of any surface or subsurface waters; lower water quality; or cause or increase flooding. Together, these two legislative acts create standards for improved management of Maine's oceans, lakes, and streams.[110]

WANT MORE INFORMATION?

EPA maintains a database of model ordinances to protect local water resources. It is accessible at: <www.epa.gov/owow/nps/ordinance/buffers.htm>.

Policy 68. Reduce lot sizes through zoning and setback requirements

Much of the low-density, dispersed development apparent today is the result of zoning requirements and building codes that specify how and where growth can occur. As discussed throughout this document, communities can improve the quality of their water resources through efforts that direct development to targeted areas and encourage more compact development that consumes less land for growth. Revised zoning and setback requirements are one way to achieve these goals.

Density bonuses encourage more growth on less land, reducing the total level of imperviousness for a community—just like guidelines that permit buildings to be constructed with smaller setbacks or less parking. Zoning codes, subdivision standards, and setback requirements all directly impact the amount of land that will be consumed by specifying minimum lot size. Communities can provide more choices to residents—and achieve water quality benefits—by revising zoning codes and subdivision standards. This action will allow development on smaller lots and lower the requirements for the distance that a building must be set back from its lot line. For example, instead of requiring a minimum of a quarter-acre for residential lots, as many current codes do, new codes could allow development on smaller lots or more units to be built on a quarter-acre parcel. Reduced setback requirements for front, side, and rear yards allow homes and commercial buildings to be built closer together and leads to shorter driveway and roadway lengths to reduce total imperviousness.

Shared driveways are another mechanism to reduce lot size while not compromising on living space.

Photo courtesy of the NEMO program and the University of Connecticut.

WANT MORE INFORMATION?

A forthcoming revision to *Parking Alternatives: Making Way for Urban Infill and Brownfields Redevelopment* expands on how localities can balance parking with broader community goals with more case studies and new proven techniques. This summer 2004 publication, *Parking Spaces/ Community Places: Finding the Balance through Smart Growth Solutions*, will available at: <www.epa.gov/ smartgrowth/ publications.htm>.

The current version can be accessed at: <www.smart growth.org/pdf/PRKGDE04. pdf>.

Market Common, a mixed-use development in Arlington, Virginia, has reduced parking requirements because of its proximity to transit and surrounding neighborhoods.

Photo courtesy of U.S. EPA.

Policy 69. Minimize parking requirements

Parking lots are a highly visible and significant share of a community's impervious surface cover; they are sizable contributors to stormwater runoff. The size and design of parking lots are currently dictated by a combination of zoning and building regulations implemented by localities, building features required by lenders, and the conventional practices of builders and developers. Communities can directly encourage smaller and more structured parking that reduces imperviousness through revised parking requirements and other supportive policies, and indirectly through education of developers and lenders.

A revised approach to parking can result in a number of water quality benefits. First, smaller parking lots and structured parking can significantly reduce the extent of imperviousness on a building site. This approach reduces the total footprint of a development, allowing more of the site to remain undeveloped or capable of absorbing additional, compact growth. Consequently, pressure to develop undisturbed land for new development is lessened, and more pervious surface is retained. In addition, a smaller parking footprint reduces the area on which pollutants can be deposited and stormwater collected, thereby reducing polluted runoff.

Also, allowing on-street parking can reduce the need for parking lots and improve walkability by helping to calm passing traffic. Montgomery County, Maryland, encourages structured parking by charging a special parking assessment on new development near the Bethesda Metro station; the money collected supports the construction and maintenance of public, multi-story parking lots in the area. The county's approach to privately constructed parking lots for offices is designed to support the use of transit, thus reducing overall parking need. The county also provides carpool and vanpool spaces in specific facilities to encourage ridesharing and tries to minimize the use of land devoted to parking by encouraging the mixed-use development of sites.[111] Other policies, such as market pricing for parking, providing only a limited amount of parking, eliminating parking subsidies, and using shared parking, can also encourage the use of transit, ride sharing, bicycling, and walking, and help reduce the demand and need for parking. Finally, communities can require that a percentage of spaces used for overflow parking be constructed with pervious or otherwise porous materials.

Finally, communities can encourage private-sector partners, such as developers and lenders, to adopt reduced onsite parking by ensuring that public transit systems are responsive to the transportation needs of potential building users. Communities can also provide information to developers and lenders on the extent to which public transit can reduce the need for parking. Although this practice can deviate from the conventional approach by lenders and developers, thorough and well-substantiated information can encourage them to reduce the amount of onsite parking provided in both residential and commercial developments.

Practice Tip: Olympia, Washington, conducted a study of the stormwater volume benefits associated with reduced impervious surfaces in new development, redevelopment, and parking lots. The city found that reducing commercial parking acreage by 20 percent could lower the impervious surface on the site by 11 percent. The city then surveyed commercial establishments to determine whether they perceived that they would be able to reduce parking by 20 percent without affecting business. In spite of the fact that business owners did not think they had excess parking, Olympia determined that the typical occupancy rate in parking lots was only 46 to 67 percent—a level clearly supportive of a 20 percent reduction. Eighteen of 31 representative sites had less than 75 percent occupancy rates during the busiest peak hours surveyed.[112]

EDUCATION

Encouraging developers and communities to consider changes in how and where growth occurs requires widespread education on smart growth alternatives and their benefits. Through outreach, training, and information sharing on new development approaches and innovative site-level construction techniques, state and local governments and water quality practitioners can help encourage smart growth practices that improve water resources.

Policy 70. Provide resources to educate developers and local staff on LID techniques

Low impact development (LID) techniques are a natural and valuable complement to a smart growth approach to achieve water quality benefits. Because they represent a significant deviation from the standard approach to development, communities can encourage their wider use by making resources available to educate developers, local staff, and others on LID techniques.

A number of resources are available to communities to support their efforts to educate staff and private-sector citizens. EPA provided support to the Low Impact Development Center to create a number of tools for communities. For example, the LID Integrated Management Practices Standards and Specifications tool helps public-works agencies design and implement their own LID standards. Another tool, the

WANT MORE INFORMATION?

The Low Impact Development Center provides various community tools at: <www.low impactdevelopment.org/EPA03.htm>.

LID Planning Process for Urban Areas, includes guidance for urban planners and landscape architects on how to incorporate LID into master plans. The LID Training Program for Linear Transportation is an interactive training program for federal, state, and local transportation agencies. Finally, the LID Sustainable School Project includes materials to help schools implement and monitor their own LID approaches as a learning tool.

Communities can achieve significant pollution prevention benefits by combining the techniques of smart growth and LID. Improved education will ensure that both approaches are used in a complementary manner to achieve the maximum possible benefits for water quality.

Practice Tip: Cherry Creek Watershed Partners in Colorado is providing resources to educate developers and staff by hiring a "Phosphorus Broker" as a way to promote better development approaches surrounding Cherry Creek. The Phosphorus Broker will identify LID techniques (such as constructed wetlands, riparian buffers, and onsite stormwater retention techniques), encourage developers to adopt these approaches, facilitate approval in the regulatory process, coordinate outreach and education on the benefits of these approaches, and promote wider implementation of these practices. This strategy serves as a contrast to the common approach in which local regulatory compliance is assessed only after construction begins.[113]

Policy 71. Create a statewide educational program for local experts

Photo courtesy of the NEMO program and the University of Connecticut.

NEMO continues its education program for its national network of water quality and land use experts.

Statewide programs to educate local experts about new practices and techniques can build valuable support for local water quality efforts. Such programs also can serve as a way for water professionals to network and share ideas. Well-educated resident experts can help guide and support local decisionmakers on development options that will have a significant water quality impact. These educational programs can also be used to encourage more general smart growth practices and create a deeper understanding among water experts on the relationship between growth, development, and water.

Practice Tip: The state of Indiana's Planning with POWER program is based on education and outreach.[114] In Indiana, all extension agents (university-based community leaders) are voting members of local zoning commissions, and are therefore in a strong position to educate other commissioners about the impacts of development on water quality. Through this program, extension agents essentially create a technical advisory committee on natural resources and water quality, comprised of local representatives from the Natural Resources Conservation Service, the Indiana Soil Conservation District, and the Indiana Department of Environmental Management. The teams hold monthly meetings and bring technical resources into the planning and zoning process.

Policy 72. Notify homebuyers of future water availability and cost

Individuals are often not fully aware of the impacts that their personal actions have on their local watershed. For example, the cumulative purchases by homebuyers of large-lot homes have a direct and significant effect on the community's overall demand for water.

Although it is not currently a common practice, local authorities, realtors, and lenders could help raise homeowner consciousness concerning water issues by educating potential homebuyers on the probability of future water limitations. Rural communities are increasingly trying to educate potential homebuyers on the realities of rural living. For example, the Planning Department in Ottawa County, Michigan, a predominately agricultural community, created a "scratch and sniff" brochure that provides future homeowners a strong whiff of how their community smells.[115] The point was simple: we are a farming community and want to stay a farming community.

Information on state water supply projections, local growth and population estimates, and anticipated policy changes (such as higher rates for excess water use), for example, could be also provided to future homebuyers. As a result, they would be better equipped to assess the likelihood that affordable water will be available in the future, which should be an important consideration when purchasing a home. Such knowledge of future water supply issues might encourage buyers to reconsider the personal and public financial impacts of large lots and the environmental effects that could result.

WANT MORE INFORMATION?

Extension agents are university-based community educators. Originally based in land grant colleges and universities, the extension program has since been expanded to include wide-ranging programs such as growth and development, water resources, and disaster mitigation. Information on Sea Grant programs can be found at: <www.sga.seagrant.org> and information on the Land Grant programs is available at: <www.reeusda.gov/1700/statepartners/usa.htm>.

Photo courtesy of USDA NRCS.

Policy 73. Educate citizens and businesses to help protect water resources

Small efforts can have a lasting impact on water quality if many participate. Oftentimes, those interested in helping maintain or improve the quality of the water are at a loss about how to contribute to the effort. When informed about behaviors that are detrimental to the environment, many individuals and businesses are likely to want to learn what they can do to help. Creating programs, educational opportunities, and incentives for behaviors that improve water quality can make a major difference in preventing additional degradation.

Tree seedlings given to children who walk to school for a week is an excellent opportunity to educate the next generation about their environmental decisions and to enhance the beauty of their school.

Schools and local civic organizations can co-sponsor special programs on how to contribute to cleaner water. Educational opportunities can be created through formal workshops or training seminars, or informal means such as fact sheets and Web-based resources. Incentives can be offered to encourage desired behavior. For example, tree seedlings could be given to children who walk to school for a week instead of riding in a car. Special community-wide events can be organized by local governments to highlight and demonstrate the impacts of individual behavior. For example, a local government could designate a "no fertilizers" month, in which homeowners and commercial buildings agree not to use fertilizers on lawns or plants. The resulting water quality impacts could then be measured and presented to the community as evidence of their successful contributions.

Efforts to educate the public about how smart growth can improve water quality, encourage more individuals to get further involved with community planning projects, and demonstrate how water-efficient technologies and designs that impact water quality are likely to result in improved behavior. Small changes in behavior will eventually translate into higher water quality on a regional basis. With a greater understanding of their individual impacts on development, communities and residents are likely to express greater support for smart growth initiatives.

Practice Tip: Portland, Oregon, implements several programs to educate individuals and businesses about their role in water quality. Programs include initiatives to disconnect rain gutters from the storm sewer system (instead directing rainfall to absorbent flowerbeds and surfaces), promote native landscaping practices, and support community-based and K-12 projects that involve hands-on activities such as tree planting and monitoring projects on school grounds to educate children about stormwater management.[116]

Policy 74. Train teachers on smart growth issues

Due to the increased development of environmental education programs (such as "reduce, reuse, recycle") during the last few decades, many children are increasingly aware of and sensitive to environmental concerns. Few, however, have an understanding of how their communities are created and shaped, and the impacts that they, as residents, have on the environment. Municipal officials and water management districts can work with local schools to incorporate smart growth issues into their curricula.

Teachers can be supported to educate their students on these connections through programs that provide them with greater capacity and resources on the issues of watershed protection, land use and development, and the principles of smart growth. As these ideas are incorporated into school curricula, children will have access to knowledge that will enable them and their families to better protect water resources.

Practice Tip: The Southwest Florida Water Management District created Project WET (Water Education for Teachers) to help build capacity in local teachers on environmental issues. The Project WET Curriculum and Activity Guide is a collection of more than 90 innovative, interdisciplinary activities that are hands-on, easy to use, and fun. These curriculum guides are available to teachers through free workshops that prepare them to educate children in K-12 about their local watershed and how to make informed decisions about water resources. The district's *Growth and Development* newsletter for high-school students provides information on how growth and development can impact natural resources. The district also provides mini-grants for classroom projects on watersheds, water quality, and alternative sources of water and conservation.[117]

WANT MORE INFORMATION?

The Trust for Public Land's publication, *Economic Benefits of Open Space*, comprehensively details the stunning economic benefits of open space. It is available at: <www.tpl.org/tier3_cdl.cfm?content_item_id=1145&folder_id=727>.

Policy 75. Encourage information-sharing among developers concerning smart growth designs that protect water resources

Communities supportive of smart growth approaches have realized there is a market segment demanding neighborhoods with vitality and diversity—with stores, parks, and businesses within walking distance of their homes. Often one barrier to building better communities is the lack of awareness from the development community. Some developers have recognized this growing market segment, in part because developments with smart growth characteristics command a market premium, yet some developers are still unaware of how to address the permitting, construction, and design issues that many smart growth developments face.

To address this barrier, more developers with a working knowledge of smart growth approaches are needed. Information-sharing among developers, through venues such as the National Association of Home Builders, about their experiences with smart growth can be a step toward meeting these needs. Because developers have intimate knowledge of the development process, they can provide valuable information on how to implement many of the ideas discussed in this section. Developers can therefore be strong advocates for techniques that protect water quality, save them money, and build better neighborhoods.

Practice Tip: The Builders for the Bay project is a unique partnership between the development and environmental communities. The Center for Watershed Protection, the Alliance for the Chesapeake Bay, and the National Association of Home Builders have agreed to hold local roundtables in the Chesapeake Bay watershed to help local jurisdictions incorporate more environmentally sensitive site designs into existing subdivision codes and ordinances. Currently, many localities require a special exception process for developers to utilize these techniques. Adoption of the regulations developed through these roundtables would provide more flexibility in the development process, help preserve natural areas, reduce stormwater runoff, and achieve cost savings. Roundtable participants include local government planning and zoning departments, watershed organizations, developers, landowners, and other community stakeholders.[118]

[85]National Stormwater Best Management Practices Database. <www.bmpdatabase.org>. Also, Center for Watershed Protection. The Stormwater Manager's Resource Center. <www.stormwatercenter.net>.

[86]North Carolina Wildlife Resources Commission. August 2002. *Guidance Memorandum to Address and Mitigate Secondary and Cumulative Impacts to Aquatic and Terrestrial Wildlife Resources and Water Quality*. <216.27.49.98/pg07_wildlifespeciescon/ pg7c3_impacts.pdf>.

[87]The Green Roof Infrastructure Monitor. 2001. "Portland Provides Incentives for Green Roof Implementation." *The Green Roof Infrastructure Monitor*. 3:1. <www.greenroofs.ca/grhcc/GRIM-Spring2001.pdf>.

[88]Florida Departments of Community Affairs and Environmental Protection. November 2002. *Protecting Florida's Springs: Land Use Planning Strategies and Best Management Practices*. <www.dca.state.fl.us/fdcp/DCp/ publications/springsmanual.pdf>.

[89]Jackson Meadow. Home page. <www.jacksonmeadow.com>.

[90]Maryland Department of the Environment. *Maryland Stormwater Design Manual, Volumes I & II* (Effective October 2000). <www.mde.state.md.us/Programs/ WaterPrograms/SedimentandStormwater/ stormwater_design/index.asp>.

[91]U.S. EPA, Office of Water. "National Menu of Best Management Practices for Storm Water Phase II." National Pollutant Discharge Elimination System (NPDES). <cfpub.epa.gov/npdes/stormwater/ menuofbmps/menu.cfm>.

[92]U.S. EPA. Development, Community, and Environment Division. June 2003. *Minimizing the Impacts of Development on Water Quality* [Draft].

[93]Maupin, Miranda and Teresa Wagner. 2003. "Regional Facility vs. On-site Development Regulations: Increasing Flexibility and Effectiveness in Development Regulation Implementation." Presented at the National Conference on Urban Stormwater: Enhancing Programs at the Local Level. Chicago, Illinois. February 17-20, 2003. U.S. EPA, Office of Wetlands, Oceans, and Watersheds. <www.epa.gov/owow/nps/ natlstormwater03/22Maupin.pdf>.

[94]Metropolitan Government of Nashville and Davidson County. Metro Stormwater Management Manual. <www.nashville.gov/stormwater/non_discharge_ policy_directive.htm>.

[95]Brown, Whitney. 1998. *Better Site Design: A Handbook for Changing Development Rules in Your Community*. Ellicott City, MD: Center for Watershed Protection. <www.cwp.org>.

[96]U.S. EPA. Development, Community, and Environment Division. February 2003. *Using Smart Growth Policies to Help Meet Phase II Storm Water Requirements* [Draft]. Also see City of San Diego. Appendix G: LEAD Method As Proposed by the City of San Diego. <www.sdcounty.ca.gov/dpw/docs/ AppendixGLeadMethod.pdf>.

[97]Portland Development Commission. 1999. Building for the Future: Sustainable Development in Portland. <www.fish.ci.portland.or.us/pdf/pdc1.pdf>.

[98]Pennsylvania Horticultural Society. February 1999. Impact 2000 - Public Landscapes. <www.pennsylvania horticulturalsociety.org/impact2000/Feb99.html>.

[99]Chicago Department of Environment. Rooftop Gardens and Green Roofs. <www.ci.chi.il.us/Environment/ AirToxPollution/UrbanHeatIsland/heatisland6.html>.

[100]Chicago Department of Environment. City Hall Rooftop Garden. <www.ci.chi.il.us/Environment/ rooftopgarden>.

[101]Prince George's County Department of Environmental Resources. January 2000. Low Impact Development Design Strategies: An Integrated Design Approach.

[102]Puget Sound Action Team. Low Impact Development. <www.psat.wa.gov/Programs/LID.htm>.

[103]U.S. EPA. Development, Community, and Environment Division. Smart Growth Policy Database. <cfpub.epa.gov/sgpdb/sgdb.cfm>.

[104]City of Columbus. 2003. Planning Overlay, 33 Columbus City Codes, ch. 3372, §701-10.

[105]Metropolitan Council. January 2000. Model Storm Water Management Ordinance. <www.metrocouncil.org/ environment/Watershed/model_sw_ord.pdf>.

[106]Grand Traverse County Department of Public Works. Stormwater and Sediment and Erosion Control Ordinance: Operation and Maintenance. <www.epa. gov/owow/nps/ordinance/documents/D2a-Grand Traverse.pdf>.

[107]New York City Department of Environmental Protection. New York City's Water Supply System. <www.nyc.gov/html/dep/html/agreement.html>.

[108]*U.S. Water News Online.* "Proposed Denver law nurtures xeriscape growth." <www.uswaternews.com/archives/arcconserv/2proden4.html>.

[109]Florida Departments of Community Affairs and Environmental Protection. November 2002. *Protecting Florida's Springs: Land Use Planning Strategies and Best Management Practices.* <www.dca.state.fl.us/fdcp/DCp/publications/springsmanual.pdf>.

[110]Witherall, Don. Maine Department of Environmental Protection. Email communication with Lynn Richards, U.S. EPA, Office of Policy, Economics, and Innovation. June 19, 2003. Also see: <www.state.me.us/dep/blwq/stand>.

[111]Montgomery County Department of Public Works and Transportation Parking Services. Public Parking District Overview. <www.dpwt.com/parking/Overview.htm>.

[112]City of Olympia, Public Works Department. Impervious Surfaces Study. <www.ci.olympia.wa.us/publicworks/images/pdf/ipds.pdf>.

[113]Cherry Creek Stewardship Partners. Home Page. <cherry-creek.org>.

[114]Planning with POWER. Home page. <www.planningwithpower.org>.

[115]Ottawa County Planning Department. October 2003. "If You Are Thinking About Moving to the County You Want to Consider This…" Ottawa County, MI.

[116]Hottenroth, Dawn, C. February, 2003. "Using Incentives and Other Actions to Reduce Watershed Impacts from Existing Development." Presented at the National Conference on Urban Stormwater: Enhancing Programs at the Local Level. Chicago, IL. February 17-20, 2003. U.S. EPA, Office of Wetlands, Oceans, and Watersheds. <www.epa.gov/owow/nps/natlstormwater03>.

[117]Southwest Florida Water Management District. Project WET. <www.swfwmd.state.fl.us/infoed/educators/wet.htm>. (Accessed July 30, 2003).

[118]Center for Watershed Protection. Builders for the Bay. <www.cwp.org/builders_for_bay.htm>.

Appendix A: Acronyms

BMP	Best management practice
CSO	Combined sewer overflow
CWA	Clean Water Act
CWP	Center for Watershed Protection
DEP	Department of Environmental Protection
EPA	Environmental Protection Agency
FEMA	Federal Emergency Management Agency
FPA	Facility planning area
GIS	Geographic information systems
IEPA	Illinois Environmental Protection Agency
LEED	Leadership in Energy and Environmental Design (Green Building Rating System)
LID	Low Impact Development
MCM	Minimum control measures
MPO	Metropolitan planning organization
NEMO	Nonpoint Education for Municipal Officials
NOACA	Northeastern Ohio Areawide Coordinating Agency
NPDES	National Pollutant Discharge Elimination System
NRCS	Natural Resources Conservation Service
NRPA	Natural Resources Protection Act
SRF	State revolving fund
SDWA	Safe Drinking Water Act
TDR	Transfer of development rights
TMDL	Total maximum daily load
USDA	United States Department of Agriculture
USGS	U.S. Geological Survey
WET	Water Education for Teachers

Appendix B: Additional Resources

GENERAL SMART GROWTH

International City/County Management Association and Smart Growth Network. *Why Smart Growth: A Primer.* <www.epa.gov/smartgrowth/publications.htm>.

Smart Growth America. <www.smartgrowthamerica.org>.

Smart Growth Network. <www.smartgrowth.org>.

Smart Growth Network and International City/County Management Association. *Getting to Smart Growth: 100 Policies for Implementation.* <www.epa.gov/smartgrowth/publications.htm>.

Smart Growth Network and International City/County Management Association. *Getting to Smart Growth II: 100 More Policies for Implementation.* <www.smartgrowth.org/pdf/gettosg2.pdf>.

U.S. Environmental Protection Agency. Smart Growth Policy Database. <cfpub.epa.gov/sgpdb/sgdb.cfm>.

U.S. Environmental Protection Agency. Smart Growth Web site. <www.epa.gov/smartgrowth>.

BEST MANAGEMENT PRACTICES

American Forests. CITYgreen software. <www.americanforests.org/graytogreen/stormwater>.

National Stormwater Best Management Practices Database. <www.bmpdatabase.org>.

U.S. Environmental Protection Agency. Urban Stormwater Best Management Practices Study. <www.epa.gov/ost/stormwater>.

U.S. Environmental Protection Agency. National Menu of Best Management Practices for Storm Water Phase II. <cfpub.epa.gov/npdes/stormwater/menuofbmps/menu.cfm>.

BROWNFIELDS

U.S. Environmental Protection Agency. Brownfields Initiative. <www.epa.gov/docs/swerosps/bf/index.html>.

U.S. Environmental Protection Agency. Smart Growth and Brownfields Initiative. <www.epa.gov/smartgrowth>.

EDUCATION/TRAINING

Nonpoint Education for Municipal Officials. <nemo.uconn.edu>.

National Association of Conservation Districts. <www.nacdnet.org>.

Sea Grant Coastal Communities and Economies. <www.iisgcp.org/team/index.htm>.

FUNDING

U.S. Environmental Protection Agency. Potential Roles for Clean Water State Revolving Fund Programs in Smart Growth Initiatives. <www.epa.gov/smartgrowth/publications.htm>.

U.S. Environmental Protection Agency. Smart Growth Funding Resource Guide. <www.epa.gov/smartgrowth/topics/funding.htm>.

GREYFIELDS

Congress for the New Urbanism. *Greyfields into Goldfields: Dead Malls Become Living Neighborhoods*. <www.cnu.org>.

LOW IMPACT DEVELOPMENT

The Low Impact Development Center. <www.lowimpactdevelopment.org>.

Low Impact Development Urban Design Tools. <www.lid-stormwater.net>.

U.S. Environmental Protection Agency. Low Impact Development Web page. <www.epa.gov/owow/nps/lid>.

STORMWATER MANAGEMENT/UTILITIES

Pioneer Valley Planning Commission. *How to Create a Stormwater Utility*. <www.pvpc.org/docs/landuse/pubs/storm_util.pdf>.

Stormwater Manager's Resource Center. <www.stormwatercenter.net>.

U.S. Environmental Protection Agency. Model stormwater ordinances. <www.epa.gov/owow/nps/ordinance/stormwater.htm>.

World Resources Institute. Credit Trading Web site. <www.nutrientnet.org>.

WATERSHED MANAGEMENT

U.S. Environmental Protection Agency. Watersheds. <www.epa.gov/owow/watershed>.

Center for Watershed Protection. <www.cwp.org>.

Trust for Public Land. *Greenprints for Growth*. <www.tpl.org>.

Appendix C: Bibliography

Alliance for the Chesapeake Bay. 1997. "Air Pollution in the Chesapeake Bay." Baltimore, MD.

American Water Works Association. 1996. WATER:\STATS Database Summary. <www.awwa.org/Communications/h20stats/ratebar.cfm>.

Arlington County, VA. February 26, 2003 Press Release. "Arlington County One of the First in the Nation To Adopt 'Main Street' Form Based Code." Arlington County, VA. <www.co.arlington.va.us/NewsReleases/Scripts/ViewList.asp>.

Arlington County, VA. Planning Division Land Use Studies, Reports, and Other Documents. <www.co.arlington.va.us/cphd/planning/docs/index.htm>.

Arnold, C.L. and C.J. Gibbons. 1996. "Impervious Surface Coverage: The Emergence of a Key Environmental Indicator." *Journal of the American Planning Association.* 62.2: 243-258.

Arnold, Chet. Nonpoint Education for Municipal Officials. December 18, 2001. Interview by Lynn Richards, U.S. EPA, Office of Policy, Economics, and Innovation.

Beach, D. 2002. *Coastal Sprawl: The Effects of Urban Design on Aquatic Ecosystems in the United States.* Pew Oceans Commission, Arlington, VA.

Brown, Whitney. 1998. *Better Site Design: A Handbook for Changing Development Rules in Your Community.* Ellicott City, MD: Center for Watershed Protection. <www.cwp.org>.

Burchell, R.W. and D. Listokin. 1995. *Land, Infrastructure, Housing Costs and Fiscal Impacts Associated with Growth: The Literature on the Impacts of Sprawl Versus Managed Growth.* New Brunswick, NJ: Rutgers University, Center for Urban Policy Research. As summarized in *The Technological Reshaping of Metropolitan America, Office of Technology Assessment.* OTA-ETI-643.

Bureau of Environmental Services. Clean River Incentive and Discount Program Update. City of Portland, OR. <www.cleanrivers-pdx.org/get_involved/stormwater_discount.htm>.

Campaign for Sensible Growth. October 2, 2002. Press Release. "Illinois EPA Preserves Facility Planning Areas." <www.growingsensibly.org/archive/archiveDetail.asp?objectID=1150>.

Caraco, Deb. 1998. *Rapid Watershed Planning Handbook-A Comprehensive Guide for Managing Urban Watersheds.* Ellicott City, MD: Center for Watershed Protection.

Center for Watershed Protection. About Builders for the Bay. <www.cwp.org/builders_for_bay.htm>.

Center for Watershed Protection. Aquatic Buffers. <www.cwp.org/aquatic_buffers.htm>.

Center for Watershed Protection. Home page. <www.cwp.org>.

Center for Watershed Protection. The Stormwater Manager's Resource Center. <www.stormwatercenter.net>.

Charles River Watershed Association. Home page. <www.crwa.org>.

Charles River Watershed Association. Water Resource Planning for Environmental Zoning: Sustaining Resources through Growth Management. <www.epa.gov/region01/ra/sprawl/grants1999.html#10>.

Cherry Creek Basin Water Quality Authority. Home page. <www.cherrycreekbasin.org>.

Cherry Creek Stewardship Partners. Home Page. <cherry-creek.org>.

Chesapeake Bay Program. A Glance: Bay Water Quality Restoration. <www.chesapeakebay.net/restoringwater.htm>.

Chicago Department of Environment. City Hall Rooftop Garden. <www.ci.chi.il.us/Environment/rooftopgarden>.

Chicago Department of Environment. Rooftop Gardens and Green Roofs. <www.ci.chi.il.us/Environment/AirToxPollution/UrbanHeatIsland/heatisland6.html>.

City of Boca Raton, FL. Downtown Redevelopment. <www.ci.boca-raton.fl.us/econ/downtown.cfm>.

City of Charlotte and Mecklenburg County Government. Charlotte-Mecklenburg Utilities. <www.charmeck.org/Departments/Utilities/Home.htm>.

City of Columbus. 2003. Planning Overlay, 33 Columbus City Codes, ch. 3372, §701-10.

City of Columbus Division of Sewerage and Drainage. November 3, 2000. Facilities Plan Update. <ci.columbus.oh.us/plan>.

City of Griffin Stormwater Department. Stormwater Utility. <www.griffinstorm.com/StormwaterUtility.htm>.

City of Indianapolis Department of Public Works. October 22, 2001. Stormwater Credit Manual. <www.indychamber.com/BusinessAdvocacy/watercredit.html>.

City of Olympia, Public Works Department. Impervious Surfaces Study. <www.ci.olympia.wa.us/publicworks/images/pdf/ipds.pdf>.

City of San Diego. Appendix G: LEAD Method as Proposed by the City of San Diego. <www.sdcounty.ca.gov/dpw/docs/AppendixGLeadMethod.pdf>.

City of Santa Monica Environmental & Public Works Management. Home page. <epwm.santa-monica.org/epwm>.

Clean Water Management Trust Fund. North Carolina Clean Water Management Trust Fund. <www.cwmtf.net>.

Cooper, Carry. Projects—Mizner Park. <www.coopercarry.com/4/4e1c.html>.

Deason, Jonathan, et al. September 2001. *Public Policies and Private Decisions Affecting the Redevelopment of Brownfields: An Analysis of Critical Factors, Relative Weights and Area Differentials.* Prepared for U.S. EPA, Office of Solid Waste and Emergency Response. Washington, D.C.: The George Washington University. <www.gwu.edu/~eem/Brownfields/project_report/report.htm>.

Denver Regional Council of Governments. "Local governments united on guiding growth: Mile High Compact signed." *Regional Report.* <www.drcog.org/pub_news/releases/RR%20Mile%20High%20Compact.pdf>.

Doll, A., P. Scodari, and G. Lindsey. 1999. "Credits as Economic Incentives for On-Site Stormwater Management: Issues and Examples." *National Conference on Retrofit Opportunities for Water Resource Protection in Urban Environments: Proceedings.* Cincinnati, Ohio: U.S. EPA. EPA/625/R-99/002.

Downing, Bob. May 27, 2002. "Stream Rules Cause Ripples." *The Beacon Journal.* Page E1.

Duany Plater-Zyberk. Transect. <www.dpz.com/transect.htm>.

Ebert, Karl. August 2002. "New stormwater rules eyed: Proposal calls for easing of regulations in city redevelopment areas." *Oshkosh Northwestern.* Oshkosh, WI: Northwestern.

Engdahl, J. 1999. *Impacts of Residential Construction on Water Quality and Quantity in Connecticut.* Storrs, CT: University of Connecticut. <www.canr.uconn.edu/jordancove>.

Environmental Trading Network. Trading Programs. <www.envtn.org/programs/programs.htm>.

Environmental Trading Network Web site. <www.gltn.org/programs/cherry/cherry.htm>.

Fairfax County Public Works and Environmental Services. 2001. *2001 Public Facilities Manual.* Fairfax County, VA.

Federal Transit Administration. New Starts. <www.fta.dot.gov/library/policy/ns/whatisns.htm>.

Florida Department of Community Affairs, Division of Community Planning. Rural Land Stewardship Areas Program. <www.dca.state.fl.us/fdcp/dcp/RuralLandStewardship/rural_lands_stewardship_areas_pr.htm>.

Florida Departments of Community Affairs and Environmental Protection. November 2002. *Protecting Florida's Springs: Land Use Planning Strategies and Best Management Practices.* <www.dca.state.fl.us/fdcp/DCp/publications/springsmanual.pdf>.

Florida Yards and Neighborhoods. Home page. University of Florida. <hort.ufl.edu/fyn/object.htm>.

Ford Motor Company. Rouge Renovation. <www.ford.com/en/goodWorks/environment/cleanerManufacturing/rougeRenovation.htm>.

Fort Worth Water Department. September 2002. Recommendation of the 2002 Informal Water and Wastewater Retail Rate Advisory Committee. <www.fortworthgov.org/water/Reports/2002ratereportEnglish.pdf>.

Frederick County Roundtable. Recommended Model Development Principles for Frederick County, MD. Center for Watershed Protection. <www.cwp.org/Frederick.pdf>

Gallagher, Megan. April 18, 2003. "New Land Trust Hosts Conservation Easement Workshop." *Eastern Shore Bulletin.* <www.billybarton.biz/bulletin/april18_2003.htm>.

Graham, Maureen. August 20, 2002. "Developers drawn into well troubles." *The Philadelphia Inquirer.* <www.philly.com/mld/inquirer/2002/08/20/news/local/states/new_jersey/counties/gloucester_county/3899875.htm>.

Grand Traverse County Department of Public Works. Stormwater and Sediment and Erosion Control Ordinance: Operation and Maintenance. <www.epa.gov/owow/nps/ordinance/documents/D2a-GrandTraverse.pdf>.

Greenroofs.com. North American Case Studies. The Greenroof Industry Resource Portal. <www.greenroofs.com/north_american_cases.htm>.

Henry, Natalie. June 13, 2003. "SoCal Counties Combining Habitat Conservation Plans with Clean Water Permits." *Land Letter*. E&E Publishing, LLC. <www.eenews.net/Landletter.htm>.

Hernandez, Bobby. U.S. EPA Region 6. May 20, 2002. Interview by Lynn Richards, U.S. EPA, Office of Policy, Economics, and Innovation.

Hillsborough County, FL. Hillsborough Greenways Program. <www.hillsboroughcounty.org/pgm/community/greenways_program.html>.

Hopper, Kim, ed. 2002. *Local Greenprinting for Growth: Using Land Conservation to Guide Growth and Preserve the Character of Our Communities, Volume 1*. San Francisco, CA: Trust for Public Land and the National Association of Counties.

Hottenroth, Dawn, C. February, 2003. "Using Incentives and Other Actions to Reduce Watershed Impacts from Existing Development." Presented at the National Conference on Urban Stormwater: Enhancing Programs at the Local Level. Chicago, IL. February 17-20, 2003. U.S. EPA, Office of Wetlands, Oceans, and Watersheds. <www.epa.gov/owow/nps/natlstormwater03>.

Hunt, William F. 2002. *Stormwater BMP Cost-Effectiveness Relationships for North Carolina*. North Carolina State University.

ICMA. 1998. *Why Smart Growth: A Primer*. Washington, DC.

Illinois Environmental Protection Agency. September 25, 2003. Press Release. "Advisory Group Selected to Oversee Pilot Testing for FPA Overhaul." <www.epa.state.il.us/news-releases/2003/2003-074-fpa-overhaul.html>.

Iowa Department of Agriculture and Land Stewardship Field Services Bureau. February 10, 2003. Iowa Water Quality/Watershed Protection Project Application, Winterset Municipal Water Supply, Cedar Lake. <www.state.ia.us/epd/wtrq/wqnews/soilwater.htm>.

Iowa Water Quality Bureau. Iowa's Drinking Water State Revolving Fund (SRF) Program. <www.state.ia.us/epd/wtrsuply/srf/srf.htm>.

Jackson Meadow. Home page. <www.jacksonmeadow.com>.

Kaspersen, Janice. November/December 2000. "The Stormwater Utility: Will it Work in Your Community?" *Stormwater, The Journal for Surface Water Quality Professionals*.

Kauffman, Gerald J., Tammy Brant, and Anne Kitchell. *The Role of Impervious Cover as a Watershed Zoning and Land Use Planning Tool in the Christina River Basin of Delaware*. <www.wr.udel.edu/publications/apaimperviouspaper.pdf>.

Lacey, WA. Ordinance No. 1113. Municipal Research and Services Center of Washington. <www.mrsc.org/ords/L32o1113.aspx>.

Lawrence, Tim. Extension Agent May 17, 2002. Interview by Lynn Richards, U.S. EPA, Office of Policy, Economics, and Innovation.

Maine Department of Environmental Protection. Contractors Certified in Erosion Control Practices. <www.maine.gov/dep/blwq/training/ccec.htm>.

Maine Department of Inland Fisheries & Wildlife. Beginning with Habitat. <www.beginningwithhabitat.org>.

Maine State Legislature, Office of the Revisor of Statutes. 2003. Title 38: Waters and Navigation. Chapter 3: Protection and Improvement of Waters. <janus.state.me.us/legis/statutes/38/title38sec420-D.html>.

Maine State Planning Office. Great American Neighborhood Sewer Extension Loan Program. Program Statement. <www.state.me.us/spo/anduse/finassist/sewer.php>.

Maryland Department of the Environment. *Maryland Stormwater Design Manual, Volumes I & II* (Effective October 2000). <www.mde.state.md.us/Programs/WaterPrograms/SedimentandStormwater/stormwater_design/index.asp>.

Maryland Department of the Environment. What makes MDE's new offices so special? <www.mde.state.md.us/AboutMDE/mp_special.ASP>.

Maryland-National Capital Park & Planning Commission, Montgomery County Department of Park & Planning. Community Based Planning. <www.mc-mncppc.org/community>.

Maupin, Miranda and Teresa Wagner. 2003. "Regional Facility vs. On-site Development Regulations: Increasing Flexibility and Effectiveness in Development Regulation Implementation." Presented at the National Conference on Urban Stormwater: Enhancing Programs at the Local Level. Chicago, Illinois. February 17-20, 2003. U.S. EPA, Office of Wetlands, Oceans, and Watersheds. <www.epa.gov/owow/nps/natlstormwater03/22Maupin.pdf>.

McElfish, James M. Jr. and Susan Casey-Lefkowitz. 2001. *Smart Growth and the Clean Water Act.* Northeast-Midwest Institute. <www.nemw.org/SGCleanWater.pdf>.

McGregor, F. Robert. September 10, 2001. "Water Quality Incentives for Development Community." Memorandum to NALGEP, Cherry Creek.

Mehan, G. Tracy, III. January 7, 2003. "Watershed-Based NPDES Permitting Policy Statement." Memorandum to Water Division Directors, Regions I-X. U.S. EPA. <www.epa.gov/npdes/pubs/watershed-permitting-policy.pdf>.

Metro. 2002. Green Streets: Innovative Solutions for Stormwater and Stream Crossings. Portland, OR. <www.metro-region.org/article.cfm?ArticleID=262>.

Metropolitan Council. January 2000. Model Storm Water Management Ordinance. <www.metrocouncil.org/environment/Watershed/model_sw_ord.pdf>.

Metropolitan Government of Nashville and Davidson County. Metro Stormwater Management Manual. <www.nashville.gov/stormwater/non_discharge_policy_directive.htm>.

Metropolitan Washington Council of Governments, Metropolitan Washington Air Quality Committee. August 13, 2003. *Plan to Improve Air Quality in the Washington, DC-MD-VA Region.* <www.mwcog.org/environment/air>.

Michigan Department of Environmental Quality. Water Quality Trading. <www.michigan.gov/deq/0,1607,7-135-3313_3682_3719---,00.html>.

Miller, Brian. Purdue University May 16, 2002. Interview by Lynn Richards, U.S. EPA, Office of Policy, Economics, and Innovation.

Minnesota Land Trust. 2001. Conservation Design Portfolio: Preserving Minnesota Landscapes Through Creative Development—An Introduction. St. Paul, MN: Minnesota Land Trust.

Montgomery County Department of Environmental Protection. 2000. Stream Conditions Cumulative Impact Models for the Potomac Subregion.

Montgomery County Department of Public Works and Transportation Parking Services. Public Parking District Overview. <www.dpwt.com/parking/Overview.htm>.

Municipal Research and Services Center of Washington. Ordinances and Resolutions. <www.mrsc.org/ords/g-l/L32o1113.htm>.

National Association of Local Government Environmental Professionals, The Trust for Public Land, and Eastern Research Group. 2003. *Smart Growth for Clean Water: Helping Communities Address the Water Quality Impacts of Sprawl.*

National Center for Environmental Innovation. State Innovation Grants Solicitation. U.S. EPA. <www.epa.gov/innovation/stategrants/sig2002.htm>.

National Stormwater Best Management Practices Database. <www.bmpdatabase.org>.

National Vacant Properties Campaign. Home page. <www.vacantproperties.org>.

Nebraska Environmental Trust. Home page. <www.environmentaltrust.org/index1.htm>.

New Jersey Department of Environmental Protection. Fact Sheet for the Draft Tier B Municipal Stormwater General Permit. <www.state.nj.us/dep/dwq/pdf/tier_b_fact.pdf>.

New Jersey Department of Environmental Protection. May 19, 2003. "Land Use Management: Water Monitoring And Standards. Surface Water Quality Standards N.J.A.C. 7:9B." <www.state.nj.us/dep/rules/adoptions/042203a.pdf>.

New Jersey Department of Environmental Protection. Municipal Stormwater Regulation Program. <www.state.nj.us/dep/dwq/municstw.html>.

New Jersey Department of Environmental Protection. The Big Map Blueprint for Intelligent Growth. <www.state.nj.us/dep/antisprawl/map.html>.

New Jersey Water Supply Authority. Raritan Basin Project Description. <www.raritanbasin.org/Description.htm>.

New York City Department of Environmental Protection. New York City's Water Supply System. <www.nyc.gov/html/dep/html/agreement.html>.

Nonpoint Education for Municipal Officials. Home page. <//nemo.uconn.edu>.

Nonpoint Education for Municipal Officials. Waterford Subdivision. Jordan Cove National Urban Monitoring Project. <nemo.uconn.edu/case_studies/jordan_cove_ct_cs.htm>.

North Carolina Department of Environment and Natural Resources, Division of Water Quality. Winter 2002. "The 10/70 Development Option of the Water Supply Watershed Protection Program." *Streamlines, A Newsletter for North Carolina Water Supply Watershed Administrators.* <h2o.enr.state.nc.us/wswp/SL/back.html>.

North Carolina Floodplain Mapping Program. <www.ncfloodmaps.com/pubdocs/NCFPMPHndOut.htm>.

North Carolina Wildlife Resources Commission. August 2002. Guidance Memorandum to Address and Mitigate Secondary and Cumulative Impacts to Aquatic and Terrestrial Wildlife Resources and Water Quality. <216.27.49.98/pg07_wildlifespeciescon/pg7c3_impacts.pdf>.

Northeastern Ohio Areawide Coordinating Agency. *Clean Water 2000: 208 Water Quality Management Plan for Northeast Ohio.* <www.noaca.org/Chapter-01__6017e_.pdf>.

Openlands Project. October 2001. *Protecting Illinois' Environment through a Stronger Facility Planning Process.* <www.openlands.org/reports/FPA%20Report.pdf>.

Oregon Department of Land Conservation and Development. Oregon's Statewide Planning Goals and Guidelines (Goal 5: Natural Resources, Scenic and Historic Areas, and Open Spaces). OAR 660-015-0000(5). <www.lcd.state.or.us/goalhtml/goals.html>.

Ottawa County Planning Department. October 2003. "If You Are Thinking About Moving to the County You Want to Consider This…" Ottawa County, MI.

Overeiner, Paul. September 2, 2003. "Officials develop initiative to improve Grand River." *Jackson Citizen Patriot.*

Paulson, C.L. 1997. Testimony on the Cherry Creek Basin Water Quality Authority before the Water Quality Control Commission of the State of Colorado.

Pennsylvania Horticultural Society. February 1999. Impact 2000—Public Landscapes. <www.pennsylvaniahorticulturalsociety.org/impact2000/Feb99.html>.

Philadelphia Green. Home page. <www.pennsylvaniahorticulturalsociety.org/phlgreen/index.html>.

Planning with POWER. Home page. <www.planningwithpower.org>.

Pollard, Trip. 2001. "Greening the American Dream?" *Planning.* 67 (10): 110-116.

Portland Development Commission. 1999. Building for the Future: Sustainable Development in Portland. <www.fish.ci.portland.or.us/pdf/pdc1.pdf>.

Prince George's County Department of Environmental Resources. January 2000. *Low Impact Development Design Strategies: An Integrated Design Approach.*

Prince William County Department of Public Works. Environmental Services. <www.co.prince-william.va.us/pworks/env_services/fee.htm>.

Pryne, Eric. May 20, 2002. "20 Years' Worth of County Land?" *Seattle Times.*

Pryne, Eric. October 2, 2001. "New customers to pick up 95% of tab for new sewers." *Seattle Times.*

Puget Sound Action Team. Low Impact Development. <www.psat.wa.gov/Programs/LID.htm>.

Rhodes, Milt. North Carolina Smart Growth Alliance. February 20, 2002. Email communication with Lynn Richards, U.S. EPA, Office of Policy, Economics, and Innovation.

Rhodes, Milt. North Carolina Smart Growth Alliance. June 27, 2003. Email communication with Madelyn Carpenter, U.S. EPA, Office of Policy, Economics, and Innovation.

Rhodes, Milt. North Carolina Smart Growth Alliance. July 11, 2003. Email communication with Lynn Richards, U.S. EPA, Office of Policy, Economics, and Innovation.

Rutgers University. 2000. *The Costs and Benefits of Alternative Growth Patterns: The Impact Assessment of the New Jersey State Plan.* Center for Urban Policy and Research. New Brunswick, NJ.

Sacramento Regional County Sanitation District. Rates and Fees. <www.srcsd.com/costs.html>.

Saginaw Bay Watershed Initiative Network. Home page. <www.saginawbaywin.org>.

San Diego Metropolitan Transit Development Board. Metropolitan Transit System Fact Sheets. <www.sdcommute.com/agencies/mts/mtdb/factsheets.asp>.

Sanchez, Rene. December 23, 2001. "New California Water Law Seeks to Curb Runaway Sprawl: Big Developments Must Show Ample Supply." *Washington Post.*

Schneider, Keith. October 23, 2002. "Ford Gives River Rouge a Green Coat." *New York Times.*

Schueler, Tom. 1994. "The Importance of Imperviousness." *Watershed Protection Techniques.* 1.3: 100-111. Ellicott City, MD: The Center for Watershed Protection. <www.stormwatercenter.net/Practice/ 1-Importance%20of%20Imperviousness.pdf>.

Shapiro, Neil. 2003. "The Stranger Amongst Us: Urban Runoff, The Forgotten Local Water Resource." Presented at the National Conference on Urban Stormwater: Enhancing Programs at the Local Level in Chicago, Illinois. February 17-20, 2003. U.S. EPA, Office of Wetlands, Oceans, and Watersheds. <www.epa.gov/owow/nps/natlstormwater03>.

Shiller, Gene. Southwest Florida Water Management District. June 27, 2003. Interview by Lynn Richards, U.S. EPA, Office of Policy, Economics, and Innovation.

Sobel, Lee, *et al.* 2002. *Greyfields into Goldfields: Dead Malls Become Living Neighborhoods.* San Francisco, CA: Congress for the New Urbanism.

South Florida Community Development Coalition. Home page. <www.floridacdc.org/policy/tif2.htm>.

Southwest Florida Water Management District. Project WET. <www.swfwmd.state.fl.us/infoed/ educators/wet.htm>.

State of California Legislative Counsel. Senate Bill No. 221, Chapter 642. <www.leginfo.ca.gov/pub/ 0102/bill/sen/sb_0201-250/sb_221_bill_ 20011009_chaptered.pdf>.

State of Massachusetts Executive Office of Environmental Affairs. Community Preservation Act Web site. <commpres.env.state.ma.us/ content/cpa.asp>.

State of New Jersey. Smart Growth Infrastructure Tax Credits Web site. <www.state.nj.us/budget02/ smarttax.html>.

Summit Soil and Water Conservation District. May 29, 2002. "The Summit County Riparian Setback Ordinance: A Fact Sheet." Cuyahoga Falls, OH. <www.summitswcd.org/riparianfactsheet.pdf>.

Sussex County Division of Environmental Resources Planning. Sparta Township, NJ Standards for Individual Commercial, and Light Industrial Subsurface Sewage Disposal Systems and Groundwater Protection From Comprehensive Land Management Code. <epa.gov/OGWDW/ protect/gwpos/nj/allordnj>.

Temple University Center for Public Policy and Eastern Pennsylvania Organizing Project. 2001. *Blight Free Philadelphia: A Public-Private Strategy to Create and Enhance Neighborhood Value.* <www.temple.edu/CPP/content/reports/ BlightFreePhiladelphia.pdf>.

The Green Roof Infrastructure Monitor. 2001. "Portland Provides Incentives for Green Roof Implementation." *The Green Roof Infrastructure Monitor.* 3:1. <www.greenroofs.ca/grhcc/ GRIM-Spring2001.pdf>.

The National Center for Smart Growth Research and Education. Education & Training <www.smartgrowth.umd.edu/education/ default.htm>.

The South Carolina Coastal Conservation League, U.S. EPA, National Oceanic and Atmospheric Administration, South Carolina Department of Health and Environment, Town of Mount Pleasant. 1995. *The Belle Hall Study: Sprawl vs. Traditional Town: Environmental Implications*. South Miami, FL: Dover, Kohl, and Partners.

Trust for Public Land. 1997. *Protecting the Source: Land Conservation and the Future of America's Drinking Water*. San Francisco, CA: Trust for Public Land.

Trust for Public Land and the National Association of Counties. 2002. *Volume 1: Local Greenprinting for Growth: Using Land Conservation to Guide Growth and Preserve the Character of Our Communities*.

Trust for Public Land. Massachusetts Community Preservation Act. <www.tpl.org/tier2_rp2.cfm?folder_id=1045>.

U.S. Census Bureau, Population Division, Population Projections Program. 2000. *Annual Projections of the Total Resident Population as of July 1: Middle, Lowest, Highest, and Zero International Migration Series, 1999 to 2100*. Washington, D.C. <www.census.gov/population/www/projections/natsum-T1.html>.

U.S. Department of Agriculture, Economic Research Service. "Major land use changes in the contiguous 48 states." *Agricultural Resources and Environmental Indicators* (AREI), 1996-97. Agriculture Handbook No.712, July 1997.

U.S. Department of Agriculture, Natural Resources Conservation Service. July 2003. Urbanization and Development of Rural Land. *2001 Annual National Resources Inventory*. <www.nrcs.usda.gov/technical/land/nri01/nri01dev.html>.

U.S. Department of Agriculture, Economic Research Service, Natural Resources and Environment Division. 1997. *National Resources Inventory*.

U.S. Department of Transportation, Federal Transit Administration, Office of Planning. The Environmental Process. <www.fta.dot.gov/office/planning/ep>.

U.S. EPA. June 2003. EPA's Draft Report on the Environment Technical Document. EPA 600-R-03-050. <www.epa.gov/indicators>.

U.S. EPA, Chesapeake Bay Program Office. The Chesapeake Bay Program Office in Philadelphia. <www.epa.gov/r3chespk>.

U.S. EPA, Development, Community, and Environment Division. April 2001. "What is Smart Growth?" EPA 231-F-01-001A.

U.S. EPA, Development, Community, and Environment Division. February 2003. *Using Smart Growth Policies to Help Meet Phase II Storm Water Requirements* [Draft].

U.S. EPA, Development, Community, and Environment Division. January 2001. *Our Built and Natural Environments*. EPA 231-R-01-002.

U.S. EPA, Development, Community, and Environment Division. June 2003. *Minimizing the Impacts of Development on Water Quality* [Draft].

U.S. EPA, Development, Community, and Environment Division. Smart Growth Policy Database. <cfpub.epa.gov/sgpdb/sgdb.cfm>.

U.S. EPA, Development, Community, and Environment Division. Table of Contents, Introduction. Smart Growth Projects by Statutory Program. <www.epa.gov/smartgrowth/pdf/sg_at_work.pdf>.

U.S. EPA, Long Island Sound Office. "EPA Takes Action To Control Nitrogen Pollution In Long Island Sound." Region 2 News & Speeches. <www.epa.gov/region02/news/2001/01026.htm>.

U.S. EPA, Long Island Sound Office. Long Island Sound Study Online. <www.epa.gov/region01/eco/lis/index.htm>.

U.S. EPA, Office of Brownfields Cleanup and Redevelopment. Brownfields Cleanup and Redevelopment. <www.epa.gov/swerosps/bf/index.html>.

U.S. EPA, Office of Policy, Economics, and Innovation and Smart Growth Funding Resource Guide. Smart Growth Network. <www.smartgrowth.org/pdf/funding_resources.pdf>.

U.S. EPA, Office of Wastewater Management. October 2000. *Potential Roles for Clean Water State Revolving Fund Programs in Smart Growth Initiatives.* EPA 832-R-00-010. <www.epa.gov/owm/cwfinance/cwsrf/smartgro.pdf>.

U.S. EPA, Office of Wastewater Management. *Voluntary National Guidelines for Management of Onsite and Clustered (Decentralized) Wastewater Treatment Systems.* <www.epa.gov/OWOWM.html/mtb/decent/management.htm>.

U.S. EPA, Office of Water. Basic information: Antidegradation policy. Water Quality Standards. <www.epa.gov/waterscience/standards/about/adeg.htm>.

U.S. EPA, Office of Water. "Great Lakes Initiative." *Water Science.* <www.epa.gov/waterscience/GLI>.

U.S. EPA, Office of Water. "National Menu of Best Management Practices for Storm Water Phase II." National Pollutant Discharge Elimination System (NPDES). <cfpub.epa.gov/npdes/stormwater/menuofbmps/menu.cfm>.

U.S. EPA, Office of Water. September 1999. Stormwater Technology Fact Sheet-Bioretention. EPA 832-F-99-012. <www.epa.gov/owm/mtb/biortn.pdf>.

U.S. EPA, Office of Water. Watershed Academy Web site. <www.epa.gov/owow/watershed/wacademy/acad2000/protection/r3.html>.

U.S. EPA, Office of Wetlands, Oceans, and Watersheds. 2000. National Water Quality Inventory: 2000 Report to Congress. <www.epa.gov/305b>.

U.S. EPA, Office of Wetlands, Oceans, and Watersheds. 2003. Fact Sheet, Water Quality Trading Policy. <www.epa.gov/owow/watershed/trading/2003factsheet.pdf>.

U.S. EPA, Office of Wetlands, Oceans, and Watersheds. April 2002. Review of Statewide Watershed Management Approaches. Final Report. <www.epa.gov/owow/watershed/approaches_fr.pdf>.

U.S. EPA, Office of Wetlands, Oceans, and Watersheds. Aquatic Buffers. Model Ordinances to Protect Local Resources. <www.epa.gov/owow/nps/ordinance/buffers.htm>.

U.S. EPA, Office of Wetlands, Oceans, and Watersheds. Case Studies. Trading. <www.epa.gov/owow/watershed/hotlink.htm>.

U.S. EPA, Office of Wetlands, Oceans, and Watersheds. Coastal Zone Act Reauthorization Amendments Section 6217. Polluted Runoff (Nonpoint Source Pollution). <www.epa.gov/owow/nps/czmact.html>.

U.S. EPA, Office of Wetlands, Oceans, and Watersheds. "Draft Trading Update-December 1996, Cherry Creek Basin, Colorado." <www.epa.gov/owow/watershed/trading/cherry.htm>.

U.S. EPA, Office of Wetlands, Oceans, and Watersheds. Erosion and Sediment Control. *Model Ordinances to Protect Local Resources.* <www.epa.gov/owow/nps/ordinance/erosion.htm>.

U.S. EPA, Office of Wetlands, Oceans, and Watersheds. Fact Sheet, Water Quality Trading Policy. <www.epa.gov/owow/watershed/trading/2003factsheet.pdf>.

U.S. EPA, Office of Wetlands, Oceans, and Watersheds. January 1993. *Guidance Specifying Management Measures for Sources of Nonpoint Pollution in Coastal Waters.* EPA 840-B-93-001c.

U.S. EPA, Office of Wetlands, Oceans, and Watersheds. Miscellaneous Ordinances. Model Ordinances to Protect Local Resources. <www.epa.gov/owow/nps/ordinance/misc.htm>.

U.S. EPA, Office of Wetlands, Oceans, and Watersheds. Ordinances and Supporting Materials. *Model Ordinances to Protect Local Resources.* <www.epa.gov/owow/nps/ordinance/osm7.htm>.

U.S. EPA, Office of Wetlands, Oceans, and Watersheds. Site Development Management Measure. Polluted Runoff (Nonpoint Source Pollution). <www.epa.gov/nps/MMGI/Chapter4/ch4-2c.html>.

U.S. EPA, Office of Wetlands, Oceans, and Watersheds. Source Water Protection. Model Ordinances to Protect Local Resources. <www.epa.gov/owow/nps/ordinance/sourcewater.htm>.

U.S. EPA, Project XL. Atlantic Steel. <www.epa.gov/projectxl/atlantic>.

U.S. EPA, Region 1. Region 1 Smart Growth Web site. <www.epa.gov/region01/ra/sprawl/grants1999.html#10>.

U.S. EPA, Stormwater Program. <cfpub.epa.gov/npdes/home.cfm?program_id=6>.

U.S. Water News Online. "Proposed Denver law nurtures xeriscape growth." <www.uswaternews.com/archives/arcconserv/2proden4.html>.

Vellinga, Mary Lynne. January 31, 2002. "Sewer Fee Plan to Limit Sprawl Gains Approval." *Sacramento Bee.* <www.sacbee.com/content/news/story/1555934p-1632412c.html>.

Vermont Agency of Natural Resources, Department of Environmental Conservation, Water Quality Division. February 2001. Management of Stormwater Runoff in Vermont: Program and Policy Options. Prepared for the Vermont General Assembly. <www.anr.state.vt.us/dec/waterq/stormwaterWIP.htm>.

Vermont Department of Environmental Conservation. Home page. State of Vermont. <www.anr.state.vt.us/dec/dec.htm>.

Washington State Department of Ecology. Stormwater. State of Washington. <www.ecy.wa.gov/programs/wq/stormwater/index.html>.

Wisconsin Department of Natural Resources. Wisconsin's Sewer Service Area Planning Program. <www.dnr.state.wi.us/org/water/wm/glwsp/ssaplan>.

Witherall, Don. Maine Department of Environmental Protection. June 19, 2003. Email communication with Lynn Richards, U.S. EPA, Office of Policy, Economics, and Innovation. <www.state.me.us/dep/blwq/stand>.

www.ingramcontent.com/pod-product-compliance
Lightning Source LLC
Chambersburg PA
CBHW080642180526

45168CB00008B/3269